EAI/Springer Innovations in Communication and Computing

Series Editor

Imrich Chlamtac, European Alliance for Innovation, Ghent, Belgium

Editor's Note

The impact of information technologies is creating a new world yet not fully understood. The extent and speed of economic, life style and social changes already perceived in everyday life is hard to estimate without understanding the technological driving forces behind it. This series presents contributed volumes featuring the latest research and development in the various information engineering technologies that play a key role in this process.

The range of topics, focusing primarily on communications and computing engineering include, but are not limited to, wireless networks; mobile communication; design and learning; gaming; interaction; e-health and pervasive healthcare; energy management; smart grids; internet of things; cognitive radio networks; computation; cloud computing; ubiquitous connectivity, and in mode general smart living, smart cities, Internet of Things and more. The series publishes a combination of expanded papers selected from hosted and sponsored European Alliance for Innovation (EAI) conferences that present cutting edge, global research as well as provide new perspectives on traditional related engineering fields. This content, complemented with open calls for contribution of book titles and individual chapters, together maintain Springer's and EAI's high standards of academic excellence. The audience for the books consists of researchers, industry professionals, advanced level students as well as practitioners in related fields of activity include information and communication specialists, security experts, economists, urban planners, doctors, and in general representatives in all those walks of life affected ad contributing to the information revolution.

Indexing: This series is indexed in Scopus, Ei Compendex, and zbMATH.

About EAI

EAI is a grassroots member organization initiated through cooperation between businesses, public, private and government organizations to address the global challenges of Europe's future competitiveness and link the European Research community with its counterparts around the globe. EAI reaches out to hundreds of thousands of individual subscribers on all continents and collaborates with an institutional member base including Fortune 500 companies, government organizations, and educational institutions, provide a free research and innovation platform.

Through its open free membership model EAI promotes a new research and innovation culture based on collaboration, connectivity and recognition of excellence by community.

More information about this series at http://www.springer.com/series/15427

Satya Prakash Yadav
Bhoopesh Singh Bhati
Dharmendra Prasad Mahato • Sachin Kumar
Editors

Federated Learning for IoT Applications

EAI
RESEARCH MEETS INNOVATION

Springer

Editors
Satya Prakash Yadav
Department of Computer Science
and Engineering
G.L. Bajaj Institute of Technology
and Management (GLBITM)
Affiliated to AKTU
Lucknow, Uttar Pradesh, India

Dharmendra Prasad Mahato
National Institute of Technology Hamirpur
Himachal Pradesh, India

Bhoopesh Singh Bhati
Department of Computer Science
and Engineering, Chandigarh University
Mohali, India

Sachin Kumar
Kyungpook National University
Daegu, Korea (Republic of)

ISSN 2522-8595 ISSN 2522-8609 (electronic)
EAI/Springer Innovations in Communication and Computing
ISBN 978-3-030-85558-1 ISBN 978-3-030-85559-8 (eBook)
https://doi.org/10.1007/978-3-030-85559-8

This Springer imprint is published by the registered company Springer Nature Switzerland AG
The registered company address is: Gewerbestrasse 11, 6330 Cham, Switzerland

Preface

Federated learning has become a new research brand in the field of IoT. Today, we see gigantic amounts data being generated by end devices such as mobile phones on a daily basis. These data contain valuable information about users and their personal preferences: what websites they mostly visit, which social media apps they generally use, what types of videos they mostly watched, etc. With such valuable information, these data become the key to building better and personalized machine learning models to deliver personalized services to maximally enhance user experiences. Federated learning provides a unique way to build such personalized models without intruding users' privacy. Such a unique advantage is the key motivation to attract researchers in the IoT community to work on this new research direction.

Federated learning also opens up a brand-new computing paradigm for IoT. As compute resources inside end devices such as mobile phones are becoming increasingly powerful, especially with the emergence of IoT chipsets, IoT is moving from clouds and datacenters to end devices. Federated learning provides a privacy-preserving mechanism to effectively leverage those decentralized compute resources inside end devices to train machine learning models. Considering that there are billions of mobile devices worldwide, the compute resources accumulated from those mobile devices are way beyond the reach of the largest datacenter in the world. In this sense, federated learning has the potential to disrupt cloud computing, the dominant computing paradigm today.

This book introduces deep understanding of federated learning and Internet of Things applications. The book explains the necessity of federated learning for IoT applications. It begins by providing a comprehensive survey of state-of-the-art research on federated learning, giving the reader a general overview of the field. This book also investigates how a personalized federated learning framework is needed in cloud-edge architecture as well as in wireless-edge architecture for intelligent IoT applications. To cope with the heterogeneity issues in IoT environments, the book investigates emerging personalized federated learning methods which are able to mitigate the negative effects caused by heterogeneities in different aspects. With the power of edge computing, the requirements for fast-processing capacity and low latency in intelligent IoT applications can also be achieved. The book provides case studies of IoT-based human activity recognition to demonstrate the effectiveness of personalized federated learning for intelligent IoT applications, as well as multiple controller design and system analysis tools including model predictive control, linear matrix inequalities, and optimal control. This unique and complete co-design framework will greatly benefit researchers, graduate students, and engineers in the fields of control theory and engineering.

Knowledge Park III, Greater Noida, Uttar Pradesh, India Satya Prakash Yadav
Mohali, India Bhoopesh Singh Bhati
Himachal Pradesh, India Dharmendra Prasad Mahato
Daegu, Korea (Republic of) Sachin Kumar

Contents

Introduction to Federated Learning

Mohit Pandey, Shubhangi Pandey, and Ajit Kumar

1 Introduction

Recently, machine learning (ML)-based technologies has grown in various artificial intelligence (AI) applications, such as image recognition, medical imaging, speech recognition, pattern recognition, and natural language processing [1–3]. Day by day, the amount of data increases. To handle this, a vast amount of data machine learning-based technologies is required. This has paved way for successes of machine leaning [2, 4]. The art of deep learning has made human life easy with the use of these human- or machine-generated data. Deep learnings-based application performs automatic face recognition systems, automatic segmentation of human organs in medical images like computed tomography images and ultrasound, etc. Any deep learning-based system requires a large amount of data to achieve significant results. For instance, Facebook reported that 3.5 billion images is required for their object detection application.

Internet of things utilizes benefits of edge computing, in which high amount of data is generated. That data is not considered as a single big data but is rather distributed among many sites. For instance, a single big amount of data transmission is difficult, and of course it is not a good idea, like Earth's image taken by satellites.

M. Pandey (✉)
School of Computer Science and Engineering, Shri Mata Vaishno Devi University, Katra, Jammu and Kashmir, India

S. Pandey
Department of Computer Application, ABES Engineering College, Ghaziabad, UP, India

A. Kumar
School of Computer Science and Engineering Soongsil University, Dongjak-gu, Seoul, South Korea

S. P. Yadav et al. (eds.), *Federated Learning for IoT Applications*, EAI/Springer Innovations in Communication and Computing,
https://doi.org/10.1007/978-3-030-85559-8_1

Another example, in machine learning-based car applications, each car in the model wants to collaborate with other cars and control center.

The updating and sharing of data, model, and control information to all points in the system with the concern of security and efficiency is a challenging task to the exiting techniques.

The training of machine learning models requires a huge amount of data as stated above. That data is generated at various sites and collected to a central hub. At the central hub, training of models can be done. The transmission of data from various sites takes lots of networking infrastructure and transmission delay. In another way, machine learning model can develop the same amount of data without transferring data to the central hub. At each site, the model can train locally with their data and share their local model to the central hub to merge a global model. This idea makes existence of federated learning possible.

Federated learning (FL) is a new emerging technology [5] that has become very popular among researchers to investigate its power and applications [6, 7]. FL tries to answer this key question [8]: Can training of machine learning model be possible without collecting data at the central hub? The center of attention in federated learning is to facilitate collaboration, as opposed to machine learning procedures where it is not always possible [9]. Also, FL is helpful to algorithm for gaining experience of system which is always not possible in the older machine learning methodology [10, 11]. FL is found in various applications such as in IoT devices, defense applications, and smart phones. Application of FL in various fields makes it reliable as per many experiments already performed. FL is full of significant power, but is still not acceptable broadly because of issues on data privacy, data access, software, and hardware [12]. In this chapter, we discuss FL's technical aspects with the help of examples of FL-based architectures [13].

1.1 Federated Learning

In federated learning, training of models is performed at data source device, which means models are set nearest to the source of data for training and prediction purposes [14]. There is also a server for receiving model updates from various local sources and combining them for developing a global model. Then by using essence of distributed computing, developed global model distributed back to the source devices [15]. Distributed computing can help track the model at various devices. In those IoT sensors and mobile devices using low-cost machine learning model, federated learning is very beneficial [16]. Figure 1 shows the basic architecture of federated learning.

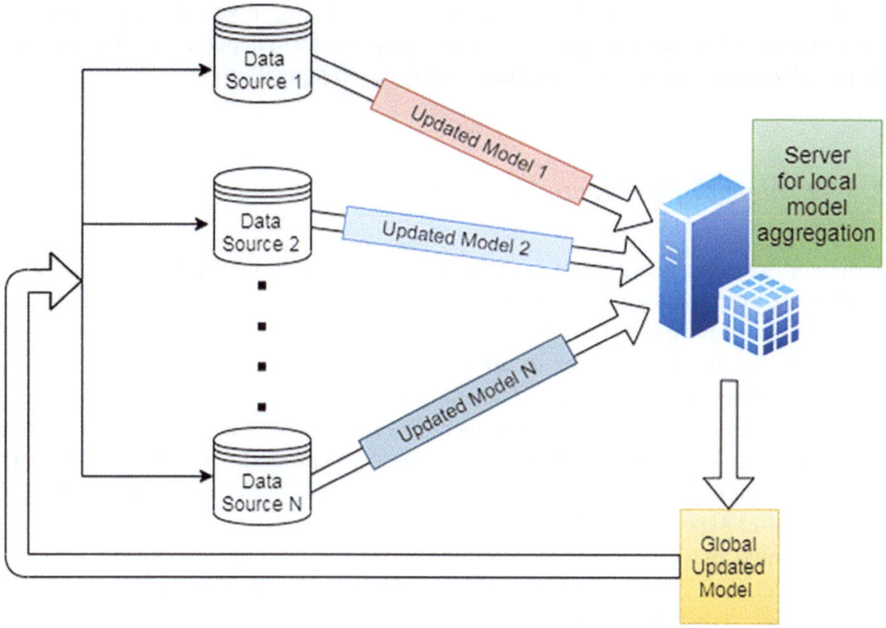

Fig. 1 Basic architecture of federated learning

1.2 Example of Application of Federated Learning

FL is applicable in various fields, and many FL-based applications have been developed most especially for the healthcare sector [17]. The basic FL architecture as shown in Fig. 1 is applied in healthcare application is showing in Fig. 2. However, it is not easy to apply FL architecture to all other areas of application. To obtain efficient results, effective learning is required. For effective learning, a large size of data is required. Data handling is also a challenging task because incoming data may be of different structure, heterogeneous contents, and various storage formats.

2 IoT and Federated Learning

Generally, federated learning is compared with deep learning, distributed learning, and parallel learning. Along with these topics, some other important new topics related to federated learning will be discussed in detail.

Machine learning algorithms when run at different parallel nodes by using distributed computing reduce training cost and make algorithm more scalable along with less memory requirement. Many applications use distributed machine learning algorithms to produce huge amount of data like in medical applications, IoT, networking, etc. In big data analytics, distributed machine learning is also very useful.

Along with the power of distributed machine learning in various applications, it has its own limitations, viz., privacy and security, platform dependency, and unavailability of standard methods to evaluate DML-based algorithms.

2.1 Concept of Federated Learning

Federated learning is a type of distributed machine learning where the training of machine learning models is performed on data available at various devices in systems like IoT devices, mobile devices, etc. [18]. The devices of the system do not send data directly to any other location, so privacy and security are handled [19]. The basic principle of federated learning is local devices can exchange updated model to create a global model at server. The server receives all parameters from various devices and averages them to generate a global model. FL ensures data privacy, less latency, and less power consumption [20]. Federated learning also enhances user experience [21].

Distributed machine learning mainly focused on parallel computing, whereas federated learning focused on the training of the model at the device where the data is generated. In distributed machine learning, the size of datasets across all nodes in the system is the same. But in federated learning, datasets size may not be the same at various nodes [22]. Nodes in distributed machine learning are data hub and establish communication using various technologies, and by which, they are able to access powerful computational resources. But devices of nodes in FL depend on battery, and communication is performed using Wi-Fi and cellular networks. However, it is not reliable, and occurrence of dropout from the network is high.

2.2 Internet of Things (IoT)

By combining two words, Internet and things, a new phrase "Internet of things" is formulated which is also known as IoT.

The Internet is a universal communication system of interconnected computer networks that use a standard Internet protocol suite (TCP/IP) to serve worldwide. It is a network of millions of businesses, private, public, government, and academic, whose connectivity is provided by a broad array of electronic, wireless, and optical networking technologies [23].

The "things" that are referred to can be any living or nonliving object. Other than the daily use of electronic devices, "things" can be any object such as apparels, foods, and furniture; materials, parts, and equipment; merchandise and specialized items; landmarks, monuments, and works of art; and all the miscellany of commerce, culture, and sophistication [24].

The Internet of Things can be considered as a global network of every and each object of the planet even beyond the planet, which allows communication between humans (H) and things (T) in various ways, viz., H2H, H2T, and T2T, by providing unique identity to each and every object [25]. That means IoT is a technology where things can communicate with the help of the Internet.

Internet of things (IoT) is a notion and a model that considers substantial presence in the environment of a variety of things/objects to interact with each other and cooperate with other things/objects to create new applications/services and reach common goals, which is possible through wireless and wired connections with unique addressing schemes. In this context, to create a smart world, there are enormous challenges in research and development, where both real and virtual and digital worlds meet, to create smart environments for energy, transport, cities, and many other areas [26].

As we discussed in the above section, the IoT connects billions of physical and virtual things which will produce tremendous amount of data and traffic on the network; hence, data storage capacity in plentiful amount is needed [27]. Also, due to open connectivity of objects, IoT will face many privacy and security challenges [28]. The Internet is relaying on TCP/IP protocol stack which will not sufficient to handle IoT enabled devices because the number of users in Internet is much smaller than IoT users.

Thus, the new architecture for IoT handles factors like interoperability, scalability, QoS, reliability, etc. Many researchers proposed architecture of IoT as in [27, 29]. Generally, the IOT system architecture is divided into three layers as shown in Fig. 2: the perception layer, the network layer, and the service layer (or application layer).

Perception Layer

It serves as source of data for IoT applications. By using various kinds of sensors (proximity sensor, pressure sensor, temperature sensor, accelerometer and gyroscope sensor, IR sensor, optical sensor, gas sensor, smoke sensor etc.), camera (often camera is in the category of sensor), wireless sensor network (WSN), RFID/NFC,

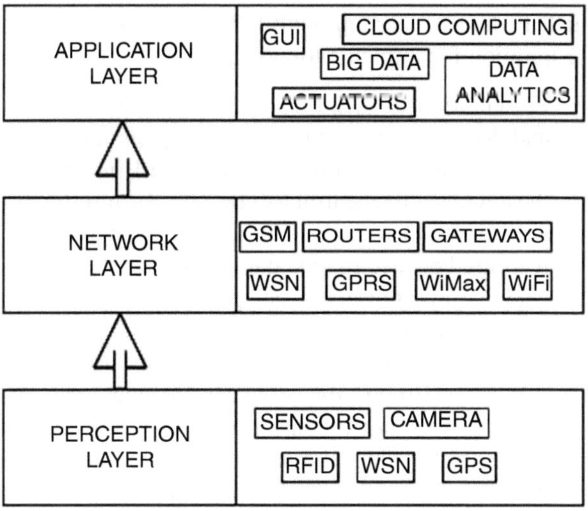

Fig. 2 Three-layer architecture of IoT

intelligent terminals, global position system (GPS), electronic data interface (EDI), etc., data about the current status of the physical world are collected.

Network Layer

This layer also called the transport layer provides infrastructure for transferring data from perception layer to above layer. This layer provides the medium between the perception layer and application layer for secure data transmission by using existing technology like WSN, mobile communication network, radio access network, and other communications equipment, such as general packet radio service (GPRS), global system for mobile communications (GSM), Wi-Fi, Ethernet, WiMax, etc. This layer also furnishes a trusted, reliable, and efficient network infrastructure platform to the upper layer and large-scale industry applications [30].

Service Layer

This layer is also known as the application layer which is divided in two sublayers, data management sublayer and application service sublayer.

The data management sublayer is responsible for managing uncertain, inaccurate, huge, and complex data by cleaning, collaboration, and restructuring. Along with this, it provides several services like Quality of Services (QoS), facility management, geomatics, etc. by cloud computing, big data, service-oriented architecture, and so on. The application service sublayer is responsible for providing graphical user interface (GUI) for enterprise application systems from bottom to upper level, such as disaster warning, logistics and supply, environmental monitoring, production management, agricultural management, and many more.

2.3 *Federated Learning and IoT*

Security, privacy, communication cost, etc. are the main obstacles in the growth of IoT. Federated learning with IoT has the potential to handle these challenges. The inspiration for implementing FL with IoT applications are summarized as follows:

- Preservation of data privacy and security: For data analytics, machine learning and deep learning are used in IoT applications. For model training, a large dataset is created by using collected data from various sources. Travel of data from the device to the central location for the training of the model makes data vulnerable to intruder attacks [31]. This problem does not happen when using federated learning because in this, data is not transferred to the central location. So federated learning preserves data privacy and security.
- Less communication cost required: IoT devices generate large amount of data continuously. This data is stored at the local node for analytical purposes and then transferred to the server. If this huge amount of data is transferred directly to the server, then it consumes very high communication cost. But in the case of federated learning, the whole data need not be communicated to the server. In this, first ML or DL where applied on the data and the only results were trans-

ferred to the server. The results are very small as compared to the whole amount of data, so communication is reduced by using federated learning.
- Scalability: Implementation of federated learning in IoT makes the deep learning procedure to scale their learning as they required.

3 Security: Federated Learning in IoT

Federated learning helps the user by protecting their valuable and confidential data because data is directly transferred to the server for training, instead of training being performed at various nodes with data generated locally. In this section, we discuss the different security mechanisms for IoT by federated learning.

3.1 Federated Machine Learning for Secure IoT

The machine learning methods are not suitable for the large amount of data as generated by IoT devices [32]. Various industries utilize a surveillance system similar to the IoT-based system. So the generated data from the various devices send to the server by results the accuracy of machine learning algorithms increases. But the main issue is the data security. For many researchers, federated ML (FML) has become very popular because it overcomes the issue of privacy by creating a local model at the devices where data is actually generated and combining these local models at a central hub. Eavesdropping and hacking are the two types of attacks in IoT [33].

Eavesdropping is performed when data is traveling on the medium. This may attack the IoT node and IoT sink and then from the IoT sink to the server communication medium.

Hacking happens at the server. Someone can find the actual location of the server by using an available stalker.

In the beginning, generated data from the devices are sent to the IoT sink. The sink is a place for collecting data from various nodes through the communication medium and also where data is encrypted before sent to the server. Then the centralized server gets the processed data from different IoT sinks and federates the data. At last, smart applications decrypt the knowledge into an understandable structure. By this, both eavesdropping and hacking are not possible, as encrypted data is present in both cases.

Data privacy is very important, and the purpose of federated learning is to develop a global machine learning model with data privacy. However, sharing of local model updates to the server makes possible the leak of personal information or some other confidential data. Some methods were suggested to preserve privacy for federated learning models from machine leanings methods.

- Differential privacy: It sets what information is sharable for third-party analysis. Sharable information under this are very basic information which consists of information about the population and some private information. Differential privacy give some confidential information and privacy protection consists of quantification of privacy loss, composition, and closure post processing, group privacy.
- Homomorphic encryption: Encryption methods were used for computation and analysis. Due to these encryption methods, malicious attacks are a difficulty for attackers [34].
- Secure multiparty computation: In this case, various parties performed computation together to ensure that information cannot leak outside of this group.

Differential privacy may not provide full data security. If the procedure by mistake considers private information as general information, then the procedure loses data protection. However, with these limitations, differential privacy has seemed to be a good privacy-preserving method compared to traditional centralized approaches.

3.2 Federated Deep Learning in IoT

In recent years, implementation of deep learning in IoT seems very popular along with edge devices. It gives real-time analytics and consumes very less resources [35]. The objective of FDL is to give IoT with enhanced power using precise deep learning. The federated deep learning framework in IoT is shown in Fig. 3. The required parameters to create the FDL model in IoT are given below:

1. FDL model: It can be on client side and server side. At client side, set the private networks consist of tuned DL model and it can précised by the global model present on the server. Précised fine-tuned models are then set down on the client side. At client side, the training of the model is performed on the data which is locally generated from edge devices. Edge devices consist of quantized and more compressed federated learning model. At the server side, the global model present at server updated by aggregating differentially gradients from each private network. All deep learning networks continuously upload gradient to the server as well as download gradient from the server [36]. The "Model Chain" was the first decentralized model which utilizes blockchain techniques to provide a privacy-preserving facility while transferring data [37]. Along with that asynchronous stochastic gradient descent is also important and used in case of a single model, in which parallel training, processing, and integration are performed.
2. Federated deep learning networking and communication: Federated deep learning makes DL models easy to run in IoT-enabled devices and enables participation of model in decision-making procedure. The decentralized deep learning process increases (i) robustness, (ii) operational efficiency, and (iii) reliability of IoT devices. In federated deep learning, two kinds of communication are possi-

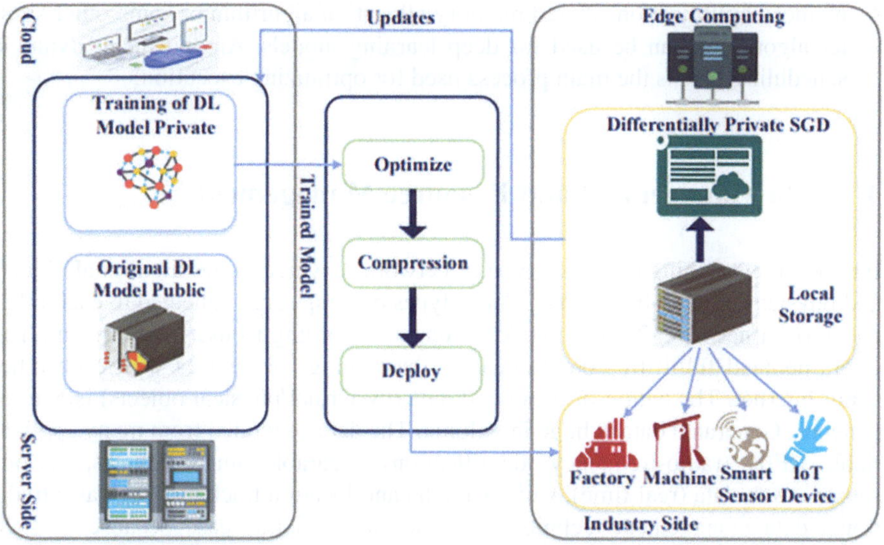

Fig. 3 Federated deep learning in IoT [40]

ble, (i) intra-communication and (ii) inter-communication. In intra-communication channel, the data is sent to every tier in the system. Between the IoT and cloud tier federated deep learning communicates. Security and privacy are handled in federated deep learning at the time of communication. The inter-communication channel is where the components in each layer establish connection and communicate with all remaining layers. They can communicate as cloud, edge, and end device. The focus of federated deep learning is to make less intra-communication and greater inter-communication, resulting in reduced overall communication cost.

3. Federated deep learning privacy and security: Many security threats are possible at the time of sharing deep learning models by federated deep learning from cloud to edge devices. Federated deep learning creates deep learning models but does not leak the information related to the data. The security challenges on the cloud consist of exchange of deep learning models which results in confidentiality and security risk. At client side, the security challenges are handled by encrypting data at the time of model training and before transferring it to the cloud.
4. Federated deep learning optimization: In the IoT, edge devices have very less computational capabilities and small memory. So, deep learning models must be small and optimized as much as possible according to the capabilities of the end devices. For example, in hardware optimization, the Graphics Processing Units (GPU) provide computation in low power consumption, so computation cost will decrease. Along with that, Google's Tensor Processing Unit [38] is also a deep learning device that utilizes fast processing of deep learning networks. For

memory optimization, shared memory allocation algorithm or some other similar algorithms can be used for deep learning models. Among them, dynamic scheduling [39] is the main process used for optimizing execution.

4 Data Management and Resource Management

Internet of everything (IoE) is a type of network in which things are digital objects and physical/virtual objects [41]. Those types of things are connected to each other and form a network. That network is created according to user requirement. The Users demand fulfill by communicating these things or devices with each other using Internet. The nature of generated data (by virtual/physical objects) is heterogeneous. Generated data is huge in volume. The data generated from many applications, RFID (radio-frequency identification), wearable smart devices, various sensors, web data (real-time), various media and location tracking. The basic traditional data management technique will not work in this case because of such unstructured, heterogeneous, big data. Various solutions were proposed by researchers, like architecture-based, middleware-based, and indexing solutions. The traditional ML is used in many smart applications, but many IoT applications are not. The flaw in real-time deployment is found because of data sensitivity and risk of sharing confidential and important information over the Internet. Resource constraints with respect to communication complexity, throughput, and latency are also found as a gap in real-time deployment of traditional ML. With capabilities of federated learning, these issues can be managed, as federated learning is mainly developed to handle data privacy. Because as already stated the federated learning cannot share data globally. In this section, we will discuss about data management and resource management.

Research interest for Internet of Everything (IoE) has increased. Internet of things (IoT) is a subset of IoE. IoT needs to handle tetrabytes of data because its devices produce such a huge amount of data. The generated data may be unstructured, structured, or semi-structured. Preserving and processing that huge amount of data is a challenging task. Many researchers handle this challenge by using various new technologies, viz., fog computing, edge computing, and cloud computing [42].

Data life cycle: The phases of the life cycle begins with pre-processing step to assess of data desired stored using fog and cloud computing [43]. The designed framework consists of edge server, proxy server, and cloud server. The time-sensitive data is placed at edge servers and processed at a fundamental level. The stored data in the edge server is transmitted to the proxy server for future purposes.

Dynamic resource provision is for energy awareness [44]. Cloud computing and edge computing work together. The drawback of this scheme is the energy consumption is high. This limitation is overcome by the three-layer architecture [45]: (i) cloud layer, (ii) edge layer, and (iii) software-defined network layer. The third layer manages network congestion, and because of this, energy consumption is reduced.

The amount of data generated through IoT-enabled devices is very huge. That large amount of data needs to be transferred from source location to the central server to be able to benefit from it, enabling heavy congestion over the Internet. This transfer needs a standard network technology and high bandwidth. The purpose of data transfer may not be achieved due to transportation delay from large data various devices make. To solve this problem, some heterogeneous architectural-based solutions were suggested by using edge computing and fog computing. However, these methods do not solve issues of security and privacy. The main problem of these types of methods is that most of the decisions are made locally, and do not consider the global data model. These challenges will be handled by federated learning.

5 Applications

Nowadays, IoT-based applications in almost every field play a significant role in our day-to-day modern life. Radio-frequency identification (RFID) technique has also become very popular in IoT-enabled application. RFID is very popular because of its very useful functions like monitoring, tracking, and supervision. Any application can opt for any function or a combination of these because these functions are interrelated to each other as shown in Fig. 4.

Generally, the term monitoring refers to the awareness of the change in the state of a system, by continuous or periodically under observation, informing whenever change in state occurs.

By continuous monitoring, you can track any person or object. Tracking continuously or sequentially records the change in location of a person or object.

These tracking records may be helpful for supervision. Supervising is deep observation of the functioning, performance, activities, or other changes in information, not only of a person but also of machine or any physical objects.

Because of these interesting functions, the application area of RFID is very broad. It captures rapidly both domestic and industrial environment; major

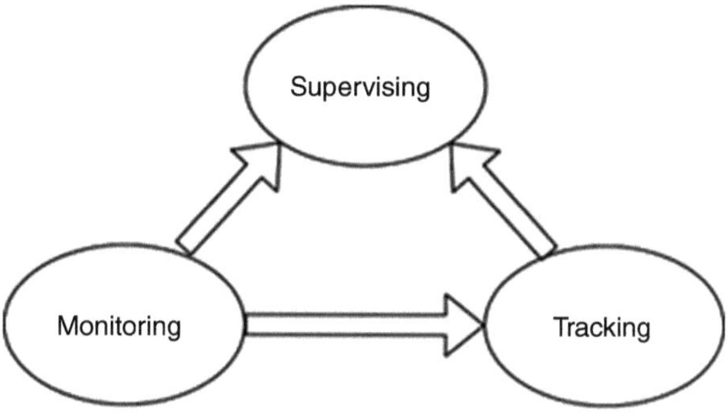

Fig. 4 Relation between various function of RFID

applications incorporate those for supply chain management (SCM), object-tracking management (OTM), and production process control.

Different fields, whether domestic or industrial, are using RFID in IoT:

- Production and manufacturing
- Agriculture management
- Payment transactions
- Transportation and retailing
- Healthcare
- Logistics and supply
- Environmental monitoring
- Disaster warning
- Warehousing
- Marine terminal operation
- Military and defense
- Distribution systems
- Other applications

In the case of manufacturing, many applications in the automotive industry have benefitted from the RFID technology, such as antitheft vehicle immobilizer devices installed for protection in many cars. RFID also play a very important role in automobile production assembly line, like production preparations, spare parts, and record management. RFID technology is not only limited to assembly line by which production cost is reduced; additionally it also offers services to users like warnings to change spare parts and ordering them and automated reminders. For both production process and end users in the automotive industry, RFID is used to provide various benefits, in terms of flexibility, visibility, added security, and traceability.

Along with evolving various types of IoT applications, and rapidly growing willingness of enterprises to utilize them, the IoT technology also emerged in fields of foundational technologies: network, software and algorithms, hardware, and data processing. The IOT is evolving with RFID and sensor technologies. From supply-chain helper to vertical-market application to ubiquitous positioning and so on, RFID technology is a very important and a basic source for IOT.

IoT with federated learning enables conversion of a traditional, old, semi-automated or manual system to a fully automated system. According to the latest GDPR regulations data generated by millions of smart sensors required privacy. So, machine learning models for smart services and products with privacy.

5.1 Federated Learning Application in Automotive Industry

Federated learning using basic essence of machine learning with additionally unlike machine learning training of the model is happens in decentralized manner [46]. The models updated with local data at various nodes i.e. data cannot share with the cloud or central server for preserving privacy and manage security. Federated learning has become very popular in IoT application with concern of privacy and security and achieved their position in various fields like medicine, transportation, smart city, entertainment, production industries, educational systems, etc. In automobile industries, the use of IoT with federated learning enables production of new smart vehicular systems by making them more automatic and able to communicate among different vehicles; the passengers and drivers of those vehicles are also able to communicate with other vehicles' occupants. Unfortunately, along with the usefulness of this new system, we cannot ignore issues like security, robustness and roadside infrastructures, etc. because the number of vehicles in this new system is increasing [47]. To handle these problems, the basic vehicular ad hoc network (VANET) [48] needs to be converted into a new concept called the federated learning-enabled Internet of vehicles (IoV) [49].

An industrial IoT model with federated learning is shown in Fig. 5. The steps followed in the model shown in the figure are given below:

1. Client node selection: Many automated system and machines make industries smarter. They have their own individual processing module and establish machine-to-machine intercommunication. An industrial IoT submodule collected data from this machine for training of model [50]. The server's role is to select the nodes for the model training. Along with data generated by the local machine in some cases data from other industries also used for making global super model. An initial central model evolves by calculating local unprocessed data of edge clients.

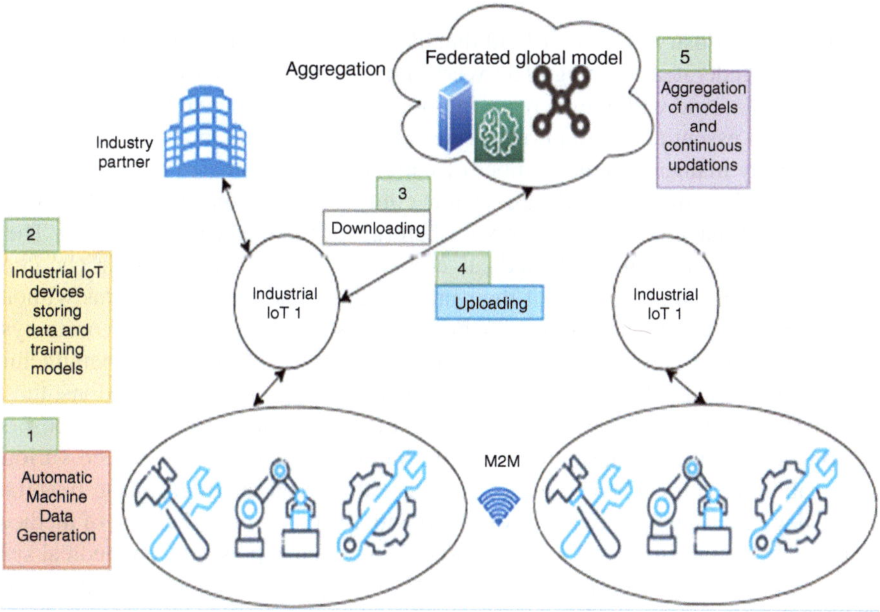

Fig. 5 A federated learning-enabled industrial IoT model [40]

2. Distribution of model: The initial model developed in the previous step is distributed to clients. The main objective of distribution is updating the model and its training. This can happen by using clients' local raw data.
3. Transferring local models: These initial models are transmitted to the central server for federated averaging.
4. Aggregation: All local models trained at various nodes are received at the central server. The server executes averaging of all received models to create a global model. That global model is more efficient and more accurate than the initial global model. Many algorithms for averaging are found in the literature, but FedAvg [50] algorithm is an excellent choice to solve averaging problem.
5. Updating model and model testing: The developed global model in the previous step is tested by the system's central task executioner. Testing is performed on the data provided by other clients (not participating in this training). Then testing results are examined if further training may be repeated, else model is distributed after updating.

The above stated steps are repeated until desired accuracy of the global model is achieved [51]. The updates produced by local entities are merged to develop a global model by using weighted average method. The main objective of industries is achieved using federated learning that is concern of efficiency.

5.2 Smart Medical Industry

The medical industry is another field that also takes benefit from machine learning methods of federated learning. The world population is increasing very fast specially in India. And with the increase in population comes a great burden on provision of good essential and required services by hospitals and the healthcare sector. To efficiently manage a large number of patients, new automated technologies are in demand [52]. In the healthcare sector, computer-aided machines and systems with IoT and machine learning are very helpful for the medical practitioner. They can provide smart tools and medical equipment. The medical sector becomes smarter with the improvement of their infrastructure and medical systems without compromising safety and taking risk prevention steps. Real-time monitoring system, fully or semi-automatic systems developed by medical equipment manufacturers, in the medical field plays a vital role. To be able to make smart decisions, these automatic systems use real-time communication with the use of various sensors, persons, different machines, etc. [53] The Medical Industry 4.0 produced various IoT-enabled devices. These devices give care to person admitted in the hospital. These devices can help patients when faced with the problem of shortage of medical staff and beds. Industrial IoT has the potential to manage pressures on medical systems with the manufacture of supplies and equipment used in hospitals [54]. Nowadays, remote medical systems provide the way for patients to be seen at the comfort of their place, i.e., no need to visit hospitals or the health center. With the remote healthcare system, the pressure on hospitals decreases but demands on healthcare services increase. Federated learning with industrial IoT systems are developed for learning persons medical reports and send to the medical expert person and they prescribed the medication for that patients while the patient is sitting at their own place. Some industrial IoT-based medical systems were developed for some particular diseases, and some systems were also developed for general routine checkup of patients. Some learning-based model for creating rehabilitation schedule in which symptoms of the particular patient is compared with the data of symptoms of old patients stored in a database and accordingly gives the best possible treatments [55]. Industrial IoT by using machine-to-machine communication provides connectivity any time and from anywhere. IoT with federated learning makes machine learning models by using local data developed by devices at local node stored at edge of the system [56]. This local model updated to global model without actually sharing the local data which provides security to manufacturing process and IoT improves productivity. In federated learning with IoT, the server collects various parameters from the local updated models and merges them to evolve a global model. The parameters of that global model are sent again to edge devices for training. This procedure is iterated until the global model achieves optimal accuracy. Thus, federated learning along with industrial IoT preserves privacy and security of sensitive data.

References

1. S. Pouyanfar, S. Sadiq, Y. Yan, et al., A survey on deep learning: Algorithms, techniques, and applications. ACM Comput. Surv. **51**(5), 1–36 (2019)
2. W.G. Hatcher, W. Yu, A survey of deep learning: Platforms, applications and emerging research trends. IEEE Access **6**(24), 411–432 (2018)
3. M. Abadi, A. Chu, I. Goodfellow, et al., Deep learning with differential privacy, in *Proc. of the ACM SIGSAC Conference on Computer and Communications Security* (2016), pp. 308–318
4. A.W. Trask, *Grokking Deep Learning* (Manning Publications, 2019)
5. P.P. Liang, et al., Think locally, act globally: Federated learning with local and global representations. arXiv preprint arXiv:2001.01523 (2020)
6. S.P. Yadav, Emotion recognition model based on facial expressions. Multimed. Tools Appl. (2021). https://doi.org/10.1007/s11042-021-10962-5
7. H. Yu, et al., A fairness-aware incentive scheme for federated learning, in *Proceedings of the AAAI/ACM Conference on AI, Ethics, and Society* (2020)
8. S. Truex, et al., A hybrid approach to privacy-preserving federated learning, in *Proceedings of the 12th ACM Workshop on Artificial Intelligence and Security* (2019)
9. B.S. Bhati, C.S. Rai, Analysis of support vector machine-based intrusion detection techniques. Arab. J. Sci. Eng. **45**(4), 2371–2383 (2020). https://doi.org/10.1007/s13369-019-03970-z. Springer
10. S.R. Pandey et al., A crowdsourcing framework for on-device federated learning. IEEE Trans. Wirel. Commun. **19**(5), 3241–3256 (2020)
11. S.P. Yadav, Vision-based detection, tracking and classification of vehicles. IEIE Trans. Smart Process. Comput. SCOPUS 9(6), 427–434 (2020). https://doi.org/10.5573/IEIESPC.2020.9.6.427. ISSN: 2287-5255

12. A. Alexander et al., An intelligent future for medical imaging: A market outlook on artificial intelligence for medical imaging. J. Am. Coll. Radiol. **17**(1), 165–170 (2020)
13. E. Bakopoulou, B. Tillman, A. Markopoulou, A federated learning approach for mobile packet classification. arXiv preprint arXiv:1907.13113 (2019)
14. T. Li, M. Sanjabi, A. Beirami, V. Smith, Fair resource allocation in federated learning. arXiv:1905.10497 (2019)
15. H. Ben Yedder, B. Cardoen, G. Hamarneh, Deep learning for biomedical image reconstruction: A survey. arXiv:2002.12351 (2020)
16. R. Doku, D.B. Rawat, C. Liu, Towards federated learning approach to determine data relevance in big data, in *Proc. IEEE 20th Int. Conf. Inf. Reuse Integr. for Data Sci. (IRI)* (Los Alamitos, CA, USA: IEEE Computer Society, 2019), pp. 184–192
17. A. Stoian, R. Ivan, I. Stoian, A. Marichescu, Current trends in medical imaging acquisition and communication, in *Proc. IEEE Int.Conf. Autom., Qual. Test., Robot*, vol 3 (IEEE, 2008), pp. 94–99
18. Q. Yang, Y. Liu, T. Chen, Y. Tong, Federated machine learning: Concept and applications. ACM Trans. Intell. Syst. Technol. **10**(2), 1–19 (2019)
19. T. Li, A.K. Sahu, A. Talwalkar, V. Smith, Federated learning: Challenges, methods, and future directions. IEEE Signal Process. Mag. **37**(3), 50–60 (2020)
20. G. Xu, H. Li, S. Liu, K. Yang, X. Lin, Verifynet: Secure and verifiable federated learning. IEEE Trans. Inf. Forensics Secur. **15**, 911–926 (2019)
21. J. Kang, Z. Xiong, D. Niyato, Y. Zou, Y. Zhang, M. Guizani, Reliable federated learning for mobile networks. IEEE Wirel. Commun. **27**(2), 72–80 (2020)
22. F. Sattler, S. Wiedemann, K.-R. Muller, W. Samek, Robust and ¨ communication-efficient federated learning from non-IID data. IEEE Trans. Neural Netw. Learn. Syst. **31**(9), 3400–3413 (2019)
23. G. Nunberg, *The Advent of the Internet: 12th April, Courses* (2012)
24. E.A. Kosmatos, N.D. Tselikas, A.C. Boucouvalas, Integrating RFIDs and smart objects into a unified internet of things architecture. Adv. Internet Things Sci. Res. **1**, 5–12 (2011)
25. R. Aggarwal, M. Lal Das, RFID Security in the context of "internet of things", in *First International Conference on Security of Internet of Things*, Kerala (2012), 17–19 August 2012, pp. 51–56
26. O. Vermesan, P. Friess (eds.), *Internet of Things: Converging Technologies for Smart Environments and Integrated Ecosystems* (River Publishers, 2013)
27. L. Tan, N. Wang, "Future internet: The internet of things, in *3rd International Conference on Advanced Computer Theory and Engineering (ICACTE)* (2010)
28. G. Gang, L. Zeyong, J. Jun, Internet of things security analysis, in *International Conference on Internet Technology and Applications (iTAP)* (2011)
29. M. Wu, T. Lu, F. Ling, J. Sun, H. Du, Research on the architecture of internet of things, in *3rd International Conference on Advanced Computer Theory and Engineering (ICACTE)* (2010)
30. Y.X. Lu, T.B. Chen, Y. Meng, Evaluation guideline system and intelligent evaluation process on the Internet of Things. Am. J. Eng. Technol. Res. **11**(9), 537–541 (2011)
31. V. Priya, I.S. Thaseen, T.R. Gadekallu, M.K. Aboudaif, E.A. Nasr, Robust attack detection approach for IIoT using ensemble classifier. Comput. Mater. Contin. **66**(3), 2457–2470 (2021)
32. L. Li, K. Ota, M. Dong, Deep learning for smart industry: Efficient manufacture inspection system with fog computing. IEEE Trans. Industr. Inform. **14**(10), 4665–4673 (2018)
33. L. Kong, X.-Y. Liu, H. Sheng, P. Zeng, G. Chen, Federated tensor mining for secure industrial internet of things. IEEE Trans. Industr. Inform. **16**(3), 2144–2153 (2019)
34. L. Kuang, L.T. Yang, J. Feng, M. Dong, Secure tensor decomposition using fully homomorphic encryption scheme. IEEE Trans. Cloud Comput. **6**(3), 868–878 (2015)
35. T. Huynh-The, C.-H. Hua, Q.-V. Pham, D.-S. Kim, MCNet: An efficient CNN architecture for robust automatic modulation classification. IEEE Commun. Lett. **24**(4), 811–815 (2020)
36. R. Shokri, V. Shmatikov, Privacy-preserving deep learning, in *Proceedings of the 22nd ACM SIGSAC Conference on Computer and Communications Security* (2015), pp. 1310–1321

37. N. Deepa, Q.-V. Pham, D.C. Nguyen, S. Bhattacharya, T.R. Gadekallu, P.K.R. Maddikunta, F. Fang, P.N. Pathirana, et al., A survey on blockchain for big data: Approaches, opportunities, and future directions. arXiv preprint arXiv:2009.00858 (2020)
38. Y.E. Wang, G.-Y. Wei, D. Brooks, Benchmarking TPU, GPU, and CPU platforms for deep learning. arXiv preprint arXiv:1907.10701 (2019)
39. H.-D. Cho, P.D.P. Engineer, K. Chung, T. Kim, *Benefits of the Big. LITTLE Architecture* (EETimes, 2012)
40. Q.-V. Pham, et al., Fusion of federated learning and industrial internet of things: A survey. arXiv preprint arXiv:2101.00798 (2021)
41. R.M. Swarna Priya, S. Bhattacharya, P.K.R. Maddikunta, S.R.K. Somayaji, K. Lakshmanna, R. Kaluri, A. Hussien, T.R. Gadekallu, Load balancing of energy cloud using wind driven and firefly algorithms in internet of everything. J. Parallel Distrib. Comput. **142**, 16–26 (2020)
42. Q.-V. Pham, H.T. Nguyen, Z. Han, W.-J. Hwang, Coalitional games for computation offloading in NOMA-enabled multi-access edge computing. IEEE Trans. Veh. Technol. **69**(2), 1982–1993 (2020)
43. J.-S. Fu, Y. Liu, H.-C. Chao, B.K. Bhargava, Z.-J. Zhang, Secure data storage and searching for industrial IoT by integrating fog computing and cloud computing. IEEE Trans. Industr. Inform. **14**(10), 4519–4528 (2018)
44. P. Borylo, A. Lason, J. Rzasa, A. Szymanski, A. Jajszczyk, Energy-aware fog and cloud interplay supported by wide area software defined networking, in *2016 IEEE International Conference on Communications (ICC)* (IEEE, 2016), pp. 1–7
45. K. Kaur, S. Garg, G.S. Aujla, N. Kumar, J.J. Rodrigues, M. Guizani, Edge computing in the industrial internet of things environment: Software-defined-networks-based edge-cloud interplay. IEEE Commun. Mag. **56**(2), 44–51 (2018)
46. B. McMahan, D. Ramage, Federated learning: Collaborative machine learning without centralized training data. Google Research Blog 3 (2017)
47. R. James et al., *The Internet of Things: A Study in Hype, Reality, Disruption, and Growth* (Raymond James US Research, Technology & Communications, Industry Report, 2014)
48. S. Zeadally, R. Hunt, Y.-S. Chen, A. Irwin, A. Hassan, Vehicular ad hoc networks (VANETS): Status, results, and challenges. Telecommun. Syst. **50**(4), 217–241 (2012)
49. Z. Du, C. Wu, T. Yoshinaga, K.-L.A. Yau, Y. Ji, J. Li, Federated learning for vehicular internet of things: Recent advances and open issues. IEEE Open J. Comput. Soc. **1**, 45–61 (2020)
50. B. McMahan, E. Moore, D. Ramage, S. Hampson, B.A. y Arcas, Communication-efficient learning of deep networks from decentralized data, in *Artificial Intelligence and Statistics*. (PMLR, 2017), pp. 1273–1282
51. T.T. Anh, N.C. Luong, D. Niyato, D.I. Kim, L.-C. Wang, Efficient training management for mobile crowd-machine learning: A deep reinforcement learning approach. IEEE Wirel. Commun. Lett. **8**(5), 1345–1348 (2019)
52. A. Challoner, G.H. Popescu, Intelligent sensing technology, smart healthcare services, and internet of medical things-based diagnosis. Am. J. Med. Res. **6**(1), 13–18 (2019)
53. S. Wang, J. Wan, D. Zhang, D. Li, C. Zhang, Towards smart factory for Industry 4.0: A self-organized multi-agent system with big data based feedback and coordination. Comput. Netw. **101**, 158–168 (2016)
54. P. Gope, T. Hwang, BSN-care: A secure IoT-based modern healthcare system using body sensor network. IEEE Sensors J. **16**(5), 1368–1376 (2015)
55. L.C.C. Gradim, M.A. Jose, D.M.C. da Cruz, R. de Deus Lopes, IoT services and applications in rehabilitation: An interdisciplinary and meta-analysis review. IEEE Trans. Neural Syst. Rehabil. Eng. **28**(9), 2043–2052 (2020)
56. S.R. Pokhrel, J. Choi, Federated learning with blockchain for autonomous vehicles: Analysis and design challenges. IEEE Trans. Commun. **68**(8), 4734–4746 (2020)

Federated Learning for IoT Devices

Deena Nath Gupta, Rajendra Kumar, and Ashwani Kumar

1 Introduction

A device in an Internet of Things (IoT) environment can be seen as its building block. A device can be a single sensor or can be made by using more than one sensor in a strategic manner. These are the entry points of data or instruction from the outside world into an IoT environment. The sensor senses its periphery and reads the required information at a regular interval. This collected information has its own value and no device wants to share its data to other devices without any deal. Also, devices use a mechanism to store only useful information and delete maximum of the sensed data because of un-usability. Federated learning (FL) enables the devices with such a mechanism by which they can perform learning on their own. Earlier, the devices had to send their data to the central processing system in order to get the learning outcome. Since learning is necessary for the end devices to make themselves compatible with the current global trends. Hence, it will be necessary for the devices to either share their data or perform the learning on their own [1–4].

Although the devices in an IoT environment are intelligent enough to make decisions by themselves; yet, a mechanism is required to update them about the new intelligent behavior. The applications under an IoT environment are highly dynamic in nature. This leads to the different information needs at different times. The devices in an IoT environment are highly constrained. Hence, they cannot be injected with many algorithms at the same time. The only possibility for a smooth functioning is to replace the old algorithm with the new one. All the participating

D. N. Gupta (✉) · R. Kumar
Jamia Millia Islamia, New Delhi, India
e-mail: rkumar1@jmi.ac.in

A. Kumar
National Institute of Technology, Patna, Bihar, India

S. P. Yadav et al. (eds.), *Federated Learning for IoT Applications*, EAI/Springer Innovations in Communication and Computing,
https://doi.org/10.1007/978-3-030-85559-8_2

devices need to be in-line so that they can get the same update. Data protection and bandwidth utilization are the main concerns with the constrained devices. The FL guarantees the protection of the data of individual devices [5].

In FL, there is no need to share one's data to the global platform. Hence, data will be within the devices only so the devices need not to worry about the misuse of their data. Also, only the gradient estimate is shared to the central authority, so the bandwidth utilization can also be improved. A new mechanism is required to enable the FL into the constrained devices. The frequency of local gradient estimates remains the open question here. There may be three possibilities to send the gradient estimate to the computing authority. The first approach is to send the gradient estimate at a predefined regular interval, second is to send the gradient estimate only when there will be any change in it, and third approach is to send the gradient estimate as and when asked by the computing authority [6].

As the learning is an integrated behavior in the federated setting, the second approach of information transfer will be most suitable here. This will also reduce the number of communication needed. Suppose there are 100 devices participating in an update. Suppose a global update is received recently. The next update will only be required when there will be prominent changes in the behavior of maximum users. If, after some time, 50 devices send a changed gradient estimate, then on the basis of these estimations and assuming another 50 inputs as previous, the global computing model will compute the next global update. The only mechanism will be required to fix the threshold for the next update [7, 8].

In this chapter, the authors will discuss some of the applications of federated learning along with challenges, threats, and their countermeasures. The details about IoT devices are presented in Sect. 2. Section 3 talks about a federated learning environment. The connection establishment between the constrained devices under a federated environment of learning is discussed in Sect. 4. The importance of actor programming is discussed in Sect. 5. Section 6 shows some challenges that come in the way of implementation of federated learning for an IoT device along with the threats and their possible countermeasures. The authors concluded their findings in Sect. 7.

2 IoT Devices

The devices in an IoT environment are highly constrained. These have very limited power as they work only with installed battery capacity. There is no source of constant power supply and also there is no way to charge the batteries of maximum devices under an IoT environment. Hence, the mechanism used in the devices should take as minimum power as possible to make them long lasting. The storage capacity of a device under an IoT environment is extremely limited. They can only store a small size program to perform a specific task. They cannot even store the intermediate results of a computation. Hence, the programmers should take care about the storage requirement while writing a program for an IoT device. The devices in an IoT environment are embedded with a tiny operating system for small computation. They can only perform computation on a predefined instruction set. There is no way to perform some other computation or some extra computation by them. The transmission range and speed of transmission are also very limited for these devices. So, an environment needs a very large number of IoT devices in order to perform well [9, 10].

The above discussed constraints make an IoT device not suitable for centralized learning. In a centralized fashion of learning, the devices have to send their sensed data to the central server all over the time. This soaks the battery power of the devices at a much faster pace because the devices need to be active all the time. The power used in transmission is very high in comparison to the internal computation. So, it will take less energy to average the inputs of say 10 measurements rather than sending all the ten measurements to the central authority. The concept of federated learning best suits here. It talks about averaging several inputs. Also there is no need to send the sensed data in federated learning makes it suitable for an IoT device. In federated learning, the devices have to send only gradient estimates of their measurements.

Suppose that a device takes "N" measurements over the time. The device will not send its data anywhere; rather it will perform averaging of the data and estimate its gradient for several periods of time.

$$\mathrm{Davg} = \left(\mathrm{D1} + \mathrm{D2} + \ldots. + \mathrm{Dn}\right) / \mathrm{n}.$$

This will reduce the number of communication required. Also, the data protection of the devices will be preserved. Suppose there are "M" number of devices taking the measurement of the same parameter from different events/locations. Each device will perform averaging separately and then forward to central computing authority whether in parallel or in serial manner.

Suppose device 1 performs averaging and then sends its gradient estimate to device 2, device 2 then performs averaging and then sends its gradient estimate to device 3, device 3 then does the same, and in this order the server will get a cumulative gradient estimate from every device in a chained fashion. This is called the vertical federated learning (VFL). The gradient estimate from one device to another is sent in a protected manner. This can be visualized by block chaining. The two cryptographic hash properties [11] need to be followed; forward secrecy and backward secrecy. In horizontal federated learning (HFL), individual devices send their gradient estimate to the global platforms individually [12–16]. Hence, there will be no need to apply cryptographic security. Instead, the injection attack might be a matter of worry. We will see later.

3 Federated Setting

The participating devices need to follow certain protocols in a federated setting. A repository is needed in all the devices to store the locally collected data. Training of a module needs a maximum amount of data. So, the first requirement for a FL device is to have a large number of entries in its repository. This is known as an example store. Every time a FL runtime is called, it accesses the example store. The FL server can call the FL runtime any time for model quality evaluation or model update.

An IoT application configures the FL runtime by providing an FL population name and registering its example store. The FL runtime, after having a handful population, contacts the FL server about its availability to run the task. It is up to the server either to provide an available task or to schedule a task in future. The FL runtime of a particular device receives the FL plan only after it has been selected by the FL server. The device can then query the example store for data requested by the plan. After getting the required data, it can compute model updates and metrics. After successful execution of an FL plan, the FL runtime reports the updates to the server. The FL runtime also cleans up the temporary resources after each successful model update. The architecture of a device in a federated setting is illustrated in Fig. 1.

The federated learning protocol includes devices and FL servers. The device takes care of FL population and the FL server works on FL tasks. The development of learning in a federated setting is threefold. In the first phase, devices connect with FL servers as and when they are ready to take an FL task. An FL task is a specific computation for an FL population, such as training to be performed with given hyper parameters, or evaluation of trained models on local device data. The FL population may be tens of thousands of devices. At a given time, the server selects only a few hundred in order to conduct a smooth operation. These devices need to be connected to the server for the whole duration. Plan, data structure, and other instructions are then sent to the participating devices. In the second phase, based on

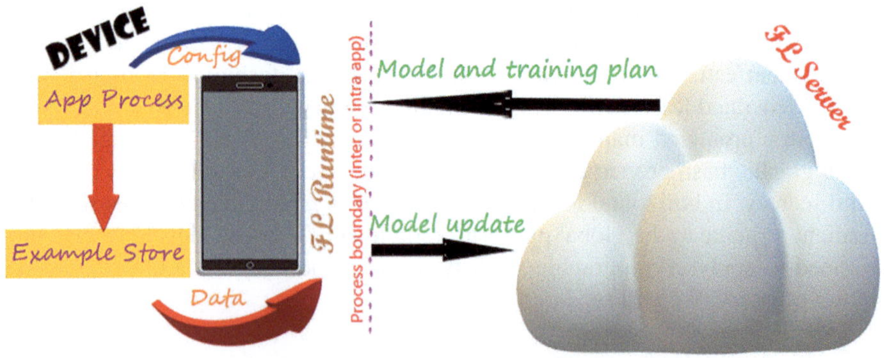

Fig. 1 Architecture of a device in a federated setting

the global parameter and device's own dataset, the device performs its local computation. In the third phase, device sends the update to the server at each FL checkpoint. These updates are incorporated by the server each time received from a device.

4 Connection Establishment

A federated setting for an IoT device maintains two actors, one database, and two states. The actors are device and server, the database is persistent storage, and the states are rejection and device or network failure. Every device opens a bidirectional stream with the server when they meet all the eligibility criteria such as battery and network/data availability. The server selects some of the connected devices for the learning based on a certain goal. Other devices are instructed to participate later in some other learning process. After device selection, the server sends FL plans and checkpoints to each participating devices. The global model is also sent to all the participating devices. The server, at the time of receiving the updates, needs an aggregation mechanism. The server can opt for a simple and secure aggregation. The server performs a federated averaging on the received updates and also instructs the devices to reconnect [17].

The success of one round depends on the received updates from enough number of participating devices. If the number of received updates are below threshold, the update will be aborted. In the federated setting, the lost communications are simply ignored. The communication can be lost by any of the two reasons; either by rejection or by network failure. The FL provides a flexible time window for devices to respond in two separate phases, i.e., selection and reporting. In the selection phase, participating device count should reach a minimum threshold before proceeding further. So the server waits for it till the timeout occurs. This will not be the same for all the rounds. Hence, the server provides a flexible time window here. Same is the case in the reporting phase, until a minimum of reports are not generated, the server has to wait.

It happens many times that the participating devices in FL learning are very low or very high. In the case of low participation, the mechanism should ensure the participation of as many devices as possible. The simultaneous connection of the devices with the server is necessary for the rate of task progress. To ensure security of the aggregation protocol, it is recommended not to allow any device to drop and no reconnection of devices if any drop happens. In the case of large participation, the check-in should be randomized. The connection of devices should be monitored as per the need. The "thundering herd" problem also needs to be taken care of. The mechanism should allow only a limited number of devices so that no devices have to wait for the participation. It should also be taken care that the same device should not be connected again and again. Similarly, it may happen that some devices find no chance to participate.

Figure 2 illustrates the complete process of aggregated global model update in a federated setting. Many devices contact the FL server for participation in learning. However, the FL server selects only a few depending on some requirement. The

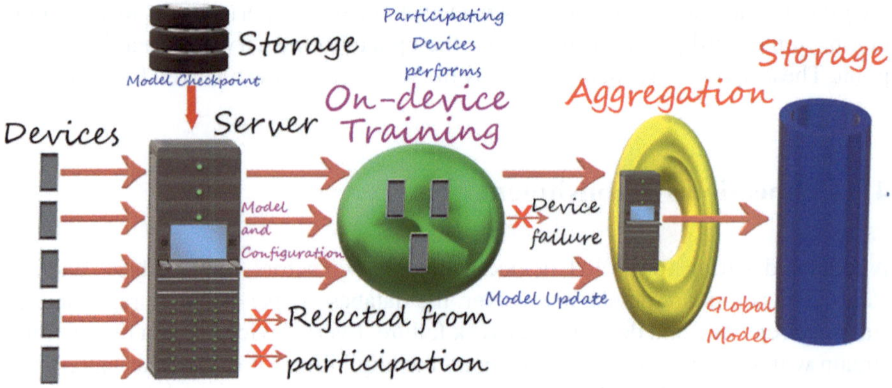

Fig. 2 Federated Learning architecture

devices not selected for the current learning process are told to connect again at some other point of time. Models and configurations are sent to the selected devices. These devices will undergo on-device training. After successful training, the devices will produce their independent update module. The aggregation server will then perform an aggregation on the received individual update modules and formulate a global update module. This global update module will then be forwarded to all the devices active in the IoT environment.

5 Actor Programming Model

Actors can serve many purposes in the distributed learning paradigm. They enable concurrent processing by message passing. A set of actors handle specific events. The same set can be used many times as per scalable condition. An agent is programmed to respond to a message, to make local decisions, to send messages to other actors, and to create more actors as per need. Dynamic resource management and load balancing can be handled with the proper placement of agents for a specific FL task. The four main actors of the architecture are coordinator, master aggregator, aggregator, and selector. The complete process is illustrated in Fig. 3.

Coordinator agent coordinates with the global platform. Multiple actors take care of different FL populations inside an IoT application. Although all other actors can access the FL population, an FL population can only be managed by the registered actor. The coordinator instructs the selectors about the maximum number of participants allowed. It then creates a master aggregator to perform the aggregation task. Master aggregator agent manages the rounds of each FL task. If the number of devices are more, the master aggregator can create more aggregators under it.

Selector agents are instructed from the coordinator about the number of devices allowed for an FL task. It then selects only the required number of devices from the pool of available devices on the basis of some mechanism. The participation of

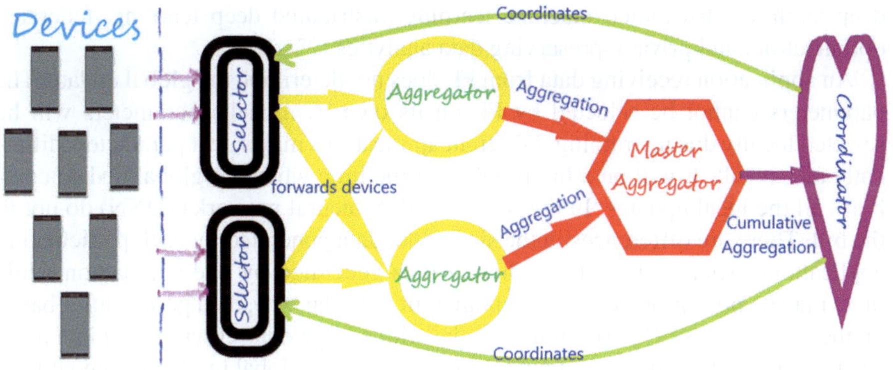

Fig. 3 Different actors in a federated setting

every device should be taken care of. No device should leave unattained and no device should overload. The devices are connected with the selector agent only. There will be no direct communication between a coordinator agent and a device. Hence, the device selection will always be independent from the coordinator and aggregator agent.

6 Challenges and Threats in FL

Users' data is often leaked by applications in order to gain financial advantage. There are also applications that exchange users' data with each other. There are many laws against this type of malpractice already in place by law enforcement agencies. The sharing of private information among companies has not been stopped completely even though many companies have been punished for leaks. In many countries, laws have been passed to restrict data collection and transactions. The country will adopt its Cyber security Law in 2017 and implement the General Provisions of Civil Law. General Data Protection Regulation (GDPR) was adopted by the European Union (EU) in 2018. In the U.S. state of California, the Consumer Privacy Act (CCPA) has recently been enacted [18–20].

IoT environments generate enormous amounts of data continuously. Safe storage is one of the most important considerations. A secure environment for data can only be found within the application itself. As a result, data must only be processed at the edge of an application. FL follows a similar methodology. Therefore, FL can be considered more appropriate for IoT environments than any other learning mechanism. Implementing FL in an IoT environment still poses some challenges. Several variations of automated learning have already been presented by the computer science community [21]. A few of them include privacy-preserving machine learning, privacy-preserving deep learning, collaborative machine learning, collaborative deep learning, distributed machine learning, distributed deep learning, federated optimization, and privacy-preserving data analytics [22–26].

An application receiving data from FL does not determine the global update. The parameters cannot be selected by FL on its own. A model's parameters will be selected locally during training. Different applications may select parameters differently as a result. It will be a big problem at the time when the global update combines all the local updates. In this scenario, deep neural networks (DNN) do not fit the bill. There are differences in the size of the data generated by each participating applications. There can be thousands of data generated by one application while there may be billions generated by another. In a similar way, data points differ based on the applications [27]. Training a model with a large dataset will result in a more accurate model than training it with a small dataset. A global model update will not be distributed equally across all participating applications. It is fair to give the big players greater access to the global updates or access early. The golden rule is to be honest. All applications cannot be guaranteed to be honest, however. An application participating in the global update may take part in unlawful activities in order to gain access to the global update. Poor practices include presenting false training results and reporting dummy data.

In vertical federated learning (VFL), nodes are connected in a serial manner. The updates of the next application depend on the updates from the previous application. Unlike in HFL, any communication loss in one application can affect all further updates. In horizontal federated learning (HFL), nodes are communicated in parallel manner. At some time, connections of multiple nodes might be broken. These nodes will not be able to send their gradient estimate to the global platform. The global modal update will be unfair for all the participating applications in this case. Unlike ML, in FL, an IoT application need not worry about the security of its data. Data generated by an application will be in its own periphery only. The only vulnerability in this case is fraudulent modal update [28]. Modal updates can be injected by malicious applications. The ecosystem will suffer in numerous ways as a result. Here are a few options to consider.

Federated computing works by separating nodes into three types namely, honest nodes, semihonest nodes, and malicious nodes. Despite training models in a way that harms ecosystems, malicious nodes train the ecosystem in a harmful way. In a federated setting, this is called a poisoning attack. As a possible solution, the trusted execution environment (TEE) is considered. Only authenticated actors are permitted to take part in trusted environments. It is required that the application be registered with the global update team prior to sending any model updates. Some malicious players will use model replacement in an environment to replace the model that has been delivered to them by the previous player with another model that is not suitable for the environment. An update of this type is made via a blockchain, using a linked-list method where all the players are connected. Various local updates are accumulated to form the global update. A backdoor attack takes place in such an arrangement [29]. Defensive distillation and adversarial training regularization are two mechanisms, which are used to safeguard the learning environment.

An equation is used to train the malicious node's model. Once the global update is over, the malicious node will be able to solve the equation in order to gain some valuable information about all the trained models and their data. Equation-solving attacks are commonly used in federated settings. A chained environment may have a node X attached, where out the model update request reaches it, this node injects its own set of modified requests into the environment. As the model is updated cumulatively by the nodes after this node X, they will take into account the modified set of requests as originally sent by the controlling authority. By responding to the requests earlier (injected by the global model), the node X can gain access. It is theorized that a malicious adversary could steal a (N + 1) dimensional linear model from a user. In formal terms, (x, h€(x)) can be expressed as a problem of solving €. By introducing the concept of k-secure, a secure two-party computation approach for vertical federated linear regression and classification is proposed to counteract the equation-solving attack.

Coordinating the gradients of a group is necessary to compute the average. In a secure system, the coordinator is assumed not to know what each participant's gradient looks like individually. The coordinator may inject a malicious model by itself and circulate the update to all participants in some cases. Working environments can be harmed by this. Extreme poisoning is what this is. "Trimmed mean" is a term proposed by Bonawitz et al. [2], in which gradients are trimmed coordinate-wise to prevent excessive poisoning [2].

7 Conclusion

In a federated setting, the devices are not asked to share their data. Devices perform all the desired computation on their own. Only an aggregate value or gradient estimate is sent to the aggregator. The communication between the device and the aggregator agent is the first point of connection where security is needed. Although devices never share the data, the adversary may find some suitable pattern from the received gradient estimate to guess the original values. The secure multiparty computation protocol is used to secure the update from individual devices by hiding them from the server. The malicious agent from the server side can get some idea from individual updates. Hence, there is a mechanism of minimum number of received updates. If the number crosses this minimum then only can a server see the cumulative sums. The selector agent at the device side can also get some insights of the device while selecting them. Hence, encryption is also needed to secure the update within the device itself. The devices need to run the secure aggregation protocol at the time of reporting only so that the updates from them can be protected if they get selected for the learning process. The protocol also needs to be robust in the case of devices dropping out more frequently.

For resource-constrained and widely distributed IoT applications, the authors discussed a distributed learning opportunity. Centralization of learning was the earlier technique. IoT applications cannot benefit from centralized learning. Data protection is one problem, and heterogeneity is another. Federated learning is a concept that protects user data. The participating devices are the only ones subjected to federated learning. Therefore, the devices need not share data with anyone. A parallel and a serial federated learning model have also been discussed. Vertical federated learning combines horizontal federated learning with device-to-device updating of models while horizontal federated learning sends the model updates in parallel. Lastly, the authors examined challenges and threats involved in applying federated learning to IoT applications. Countermeasures need to be formulated in the future to combat these threats.

References

1. Q. Yang et al., Federated Learning
2. K. Bonawitz et al., Towards Federated Learning at Scale: System Design, Feb. 2019, [Online]. Available: http://arxiv.org/abs/1902.01046
3. J. Konečný, H.B. McMahan, F.X. Yu, P. Richtárik, A.T. Suresh, D. Bacon, Federated Learning: Strategies for Improving Communication Efficiency, Oct. 2016, [Online]. Available: http://arxiv.org/abs/1610.05492
4. Q. Yang, Y. Liu, T. Chen, Y. Tong, Federated Machine Learning: Concept and Applications, Feb. 2019, [Online]. Available: http://arxiv.org/abs/1902.04885
5. F. Bonomi, R. Milito, J. Zhu, S. Addepalli, Fog computing and its role in the internet of things, *Proc. first Ed. MCC Work. Mob. cloud Comput. - MCC '12*, p. 13 (2012). https://doi.org/10.1145/2342509.2342513
6. D.N. Gupta, R. Kumar, A. Kumar, Efficient encryption techniques for data transmission through the internet of things devices, in *IoT and Cloud Computing Advancements in Vehicular Ad-Hoc Networks*, ed. by V. Jain, O. Kaiwartya, N. Singh, R. S. Rao, 1st edn., (IGI Global, Pennsylvania, 2020), pp. 203–228
7. A.R. Chowdhury, J. Mahmud, A.R.M. Kamal, M.A. Hamid, MAES: Modified advanced encryption standard for resource constraint environments, 2018 *IEEE Sensors Appl. Symp. SAS 2018 - Proc.*, vol. 2018-Janua, pp. 1–6 (2018). https://doi.org/10.1109/SAS.2018.8336747
8. Y. Qiu, M. Ma, A mutual authentication and key establishment scheme for M2M communication in 6LoWPAN networks. IEEE Trans. Ind. Informatics **12**(6), 2074–2085 (2016). https://doi.org/10.1109/TII.2016.2604681
9. D.N. Gupta, R. Kumar, Generating Random Binary Bit Sequences for Secure Communications between Constraint Devices under the IOT Environment, *INCET* (2020), pp. 1–6
10. D.N. Gupta, R. Kumar, Lightweight cryptography : an IoT perspective. Int. J. Innov. Technol. Explor. Eng. **8**(8), 700–706 (2019). [Online]. Available: https://www.ijitee.org/download/volume-8-issue-8/
11. D.N. Gupta, Sponge-based Lightweight Cryptographic Hash Functions for IoT Applications (2021), pp. 1–5
12. S.P. Yadav, K.K. Agrawal, B.S. Bhati, F. Al-Turjman, L. Mostarda, Blockchain-based cryptocurrency regulation: An overview. Comput. Econ. 0123456789 (2020). https://doi.org/10.1007/s10614-020-10050-0
13. V. Kumar, G.S. Lalotra, Blockchain-Enabled Secure Internet of Things (2020), pp. 45–55. https://doi.org/10.4018/978-1-7998-2414-5.ch003
14. T. Pizzuto, Blockchain Technology Primer (2019), pp. 140–151. https://doi.org/10.4018/978-1-7998-2414-5.ch009
15. Blockchain as an Enabler for Zero-Trust Architectures (2020), pp. 234–263. https://doi.org/10.4018/978-1-7998-3979-8.ch008

16. K.K. Sreelakshmi, A. Bhatia, A. Agrawal, Securing IoT Applications Using Blockchain (2020), pp. 56–83. https://doi.org/10.4018/978-1-7998-2414-5.ch004
17. D.N. Gupta, R. Kumar, Networking in IoT: Technologies used, security threats and possible countermeasures. Int. J. Sensors Wirel. Commun. Control. **10** (Jan. 2021). https://doi.org/10.2174/2210327910666210106090826
18. T.U. Rehman, International context of cybercrime and cyber law, Encycl. Crim. Act. Deep Web, 412–423 (2020). https://doi.org/10.4018/978-1-5225-9715-5.ch028
19. O.S.L. Tan, R.G. Vergara, R.C.W. Phan, S. Khan, N. Khan, Cybersecurity laws in Malaysia. Encycl. Crim. Act. Deep Web, 435–448 (2020). https://doi.org/10.4018/978-1-5225-9715-5.ch030
20. O.S.L. Tan, R.G. Vergara, R.C.W. Phan, S. Khan, N. Khan, Human trafficking and cyber laws in Malaysia. Encycl. Crim. Act. Deep Web, 518–532 (2020). https://doi.org/10.4018/978-1-5225-9715-5.ch035
21. D.N. Gupta, R. Kumar, Distributed key generation for secure communications between different actors in service oriented highly dense VANET, *Cloud and IoT-Based Vehicular Ad Hoc Networks*. John Wiley & Sons, Ltd (2021), pp. 221–232
22. H. Brendan McMahan, E. Moore, D. Ramage, S. Hampson, B. AgüeraAg, A. Arcas, Communication-Efficient Learning of Deep Networks from Decentralized Data (2017)
23. A.F. Skarmeta, J.L. Hernandez-Ramos, M.V. Moreno, A decentralized approach for security and privacy challenges in the Internet of Things, 2014 *IEEE World Forum Internet Things, WF-IoT* (2014), pp. 67–72. https://doi.org/10.1109/WF-IoT.2014.6803122
24. A. Da Veiga, Achieving a Security Culture (2019), pp. 72–100. https://doi.org/10.4018/978-1-5225-7847-5.ch005
25. S. Naoui, M.E. Elhdhili, L.A. Saidane, Lightweight enhanced collaborative key management scheme for smart home application, *Proc.* - 2017 *Int. Conf. High Perform. Comput. Simulation, HPCS 2017* (2017), pp. 777–784, https://doi.org/10.1109/HPCS.2017.117
26. M.A. Thomas, D.S. Abraham, D. Liu, An Assessment of Federated Machine Learning for Translational Research (2020), pp. 123–142. https://doi.org/10.4018/978-1-7998-1879-3.ch006
27. O. Vermersan, P. Friess, IoT-From Research and Innovation to Market Deployment, **6**(1) (2014)
28. V. Rajasekaran, J. Mohana, Commerce Computer Science SECURITY ISSUES WITH POSSIBLE SOLUTIONS Assistant Professor, Department of Computer Science, PSG College of Arts And Science, Coimbatore Assistant Professor, Department of Computer Science, PSG College of Arts And Science, (1), 48–50 (2015)
29. V. Venkatesh, S. Gordon, The Intersection of Cybercrime and the Blockchain (2020), pp. 676–699. https://doi.org/10.4018/978-1-7998-5728-0.ch031

Personalized Federated Learning

Kaushal Kishor

1 Introduction

Federated learning is a comparatively modern approach to education that prevents the use of structured data collection and model teaching. Data is compiled from various channels (e.g., cell devices) and processed in a central place in a typical machine learning pipeline (i.e., data center). Once all of the data is centralized, a single machine learning algorithm is trained using it. This method is known as centralized learning since the data must be transferred from the users' devices to a central computer for the model to be built and trained. Federated learning, on the other hand, involves training several machine learning models on mobile devices (known as clients) and then merging the effects into a single model that exists on a server. Thus, utilizing ground-truth evidence, a model is trained on the machines themselves, and only the trained model is exchanged with a server. The user's data is used to create machine/deep learning models in this manner, although the data stays private. In this situation, federated learning takes advantage of the users' data without sharing their personal information. The raw data resides on the users' computers and is never submitted to a data center; instead, a model is generated from the data and sent to the server. The user's data is not uploaded to the cloud by using federated learning, because there is no easy link to the data, although it is always possible to access it. In a later article, we'll speak about how anonymity is violated in federated learning.

The central principle driving the new wave of artificial intelligence is "information protection." When data protection is a big concern and we don't trust anybody to keep our information confidential, we may use federated learning to create

K. Kishor (✉)
IT Department, ABES Institute of Technology, Ghaziabad, UP, India
e-mail: kaushal.kishor@abesit.in

S. P. Yadav et al. (eds.), *Federated Learning for IoT Applications*, EAI/Springer Innovations in Communication and Computing,
https://doi.org/10.1007/978-3-030-85559-8_3

privacy-preserving AI by developing intelligent systems privately. The goal of federated learning is to move computations to results. In cases where an internationally shared model is purchased for use in places where data is stored, such as smart phones. By transferring the model to the unit, we can train the model as a whole. Anyone will engage in federated learning on their computers, either explicitly or implicitly, provided they keep this principle in mind. Edge devices, such as smart phones and IoT devices, may gain from on-chip data without the data actually leaving the device, which is particularly useful for computationally limited devices where connectivity is a bottleneck. Moving computations to data is a powerful principle for constructing any intelligent device while safeguarding the privacy of any person.

1.1 Work-Related Federal Learning with IoT Devices

More and more, the Internet of things (IoT) is growing as a functional advancement for smart environments, homes, and personal healthcare [1]. A network has been referred to as "Connected Internet, Records, and Knowledge Sharing" [2]. Adding greater cohesiveness to industrial IoT systems has strengthened their intensity and productivity; this system is great for activities such as water quality control, honeybee colony monitoring, data connectivity, spatiotemporal forecasting, and intelligent farm management. Several studies are currently being done to keep the importance of IoT technologies intact [3]. The process of connecting the virtual environment to the real world is assimilation [4]. Connecting the virtual universe to the physical world would be difficult [5]. Stuff and people will locate each other while they use this interface [6]. The aim of the Internet of things is to attach machines so they can share information and data. Smart phones, automobiles, medical supplies, and previously wired computers are examples of widely owned personal products [7]. Additionally, RFID has played a key role in the development of sensors, smart objects, and actuators [8]. However, real-world applications of machine-to-machine (M2M) [9] and vehicle-to-vehicle (V2V) [10] show the IoT's major benefits [11, 12].

Endless networks, sensing, encoding, and synchronization would be given to billions of users as part of the Internet of things (IoT) (such as cellphones and sensors). Machine learning is applied to collect intelligence from vast volumes of IoT devices to include a broad variety of intelligent services. IoT companies, autonomous automobiles, and clinics and supply chain financing are some of the many technologies in progress. Nonetheless, there are many problems that necessitate the paradigm change from "linked artifacts" to "connected knowledge" [13]. First, putting sensitive data in the cloud opens the data up for assault. Second, because of limited bandwidth, aggregating a lot of distributed data over radio channels is likely to raise network latency and generate congestion. The third thing is that many embedded systems are able to handle plenty of data and yet have unused space [14].

Distributed intelligent IoT systems will use federated machine learning [15]. Fused machine learning requires devices to store their data locally and have the data sent to an edge location-aggregation server only through training. This prohibits Internet of things (IoT) data from being exchanged with other Internet device servers or the aggregation server [16, 17]. For the best performance, a shared global model can be trained at the edge at the fleet of IoT computers. Bypassing advanced federated machine learning will minimize data transfer to the cloud to reduce learning latency. Though additional user selection is possible, frequent device connection updates from selected IoT devices will significantly reduce the system's efficiency for machine learning-based applications.

This technique utilizes the signal superposition of multi-property of federated machine learning to achieve simultaneous local model aggregation [18–20]. That can be accomplished by concurrently uploading a modified model to each framework, and having an average of such at the server. This preferably accommodates the local model changes in federated machine learning. Rather than perceiving access networks as interference, AirComp utilizes channel diversity to assist the learning phase over the access network. The efficiency of AirComp still depends on Wi-Fi requirements.

Reconfigurable Intelligent Surface (RIS) [21] has the ability to minimize Air Aggregation Error Modeling's aggregate error by reconfiguring the wireless conditions. A passive-multiple flat metasurface with software-enabled adjustable phases by exchanging all reflecting components, RIS is able to monitor the strength of incident signals to improve co-reflected signals. New technologies include edge learning, multi-input/multi-output computing [22], the usage of terahertz networking, and sparse-code multi-access [23]. We can build an intelligent IoT machine learning system using a new access strategy that helps improve communication efficiency while minimizing model complexity

2 Internet of Things Architecture

There is no widely agreed-upon framework for the Internet of things. Various scholars have suggested numerous architectures.

2.1 Architectures of Three to Five Layers

The figure in Fig. 1, referred to as a three-expanded architecture, is the most basic design. The appearance of this discovery was first noted in the very early stages of this field's study. The UI is constructed out of three components: a graphical interface, a network of interconnected systems, and an application. Aspects of the natural environment that are sensed and logged require sensors that are included in the physical layer to gather awareness. The vessel recognizes some spatial parameters

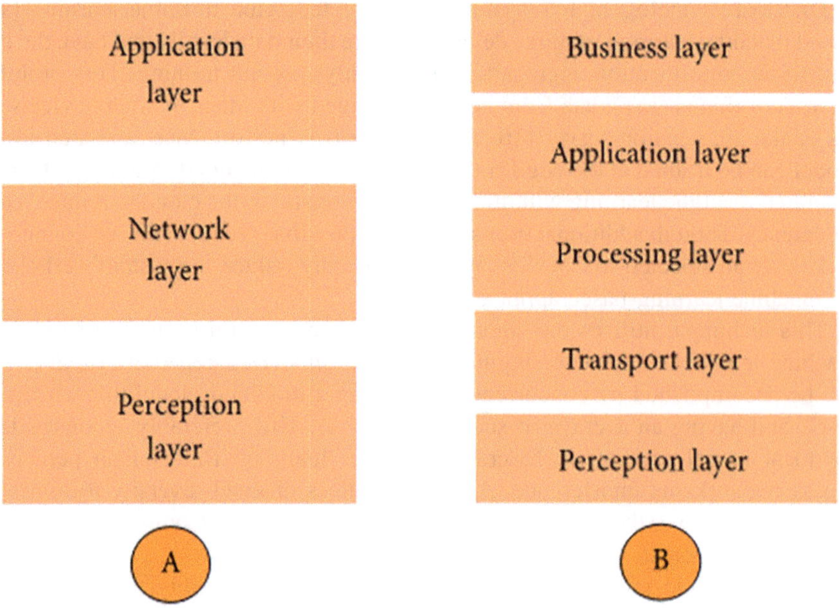

Fig. 1 Data architecture multilayered (B: five layers)

in the area and expands them (stretches them) to their limit. Both connecting to other smart objects and networking devices, allowing certain devices to connect to one another, and accepting connectivity from outside of the network are duties of the network layer. The capability of these machines to accurately extract sensor readings is also put to use in the transmission and receipt of sensor data. Applications including running programs, creating database records, and storing data are provided by the service layer. The Internet of things goes on to many purposes, such as smart homes and smart cities and, also, on top of that, often includes how they are used for well-being.

The three-layered architecture incorporates the key ideas of the Internet of things but frequently depends on additional evidence. As a result, several more layered architectures have been suggested in the literature. The five-layer design [24–27], for example, comprises the processing and market layers. Perception, transport, processing, application, and enterprise are the five levels (see Fig. 1). The three-layer architecture has three functions. The other three layers are detailed. (i) The application layer oversees all application processes using information collected from the middleware layer. (ii) In networks such as wireless, 3G, Bluetooth, and RFID, sensor data is transmitted from the transceiver to the processor layer. It is also thought of as a middleware layer. Massive volumes of data from the transport layer are collected, interpreted, and processed by it. It has the capacity to handle and offer a range of resources to the lower layers. Databases, cloud storage, and large data analytics modules are among the tools included. (iii) The business layer is in control of the whole IoT framework, including software, business and benefit structures, and the privacy of consumers. The focus of this paper does not include the company layer. As a result, we will not continue to explore it.

Another architecture suggested by Ning and Wang [28] is focused on the human brain's layers of computation. It is driven by human intellect and the capacity to perceive, feel, recall, make choices, and respond to their environments. It is made up of three pieces. Computing and storage are also located in the human brain. The second part of the body is a decentralized hierarchical network of distribution centers. The network which refers to modules and sensors are the third element.

2.2 *Architectures Centered on the Cloud and Fog*

Let's see how we can use two machine models: cloud and fog computing [29]. Additional details appear in Sect. 2.

Data provided by IoT devices is unclear, as well as the data type. A streamlined approach to data mining is used in many cloud architectures. The cloud is at the core of this cloud-centric architecture, with software on top and things running below [30]. Cloud infrastructure is prioritized because of its scalability and versatility. It provides central technology, platform, applications, and storage facilities. Developers may use the cloud to distribute their servers, applications, data processing, deep learning, and visualization resources.

Fog computing [31, 32] is a new development in device engineering in which sensors and network gateways capture and analyze data. Figure 2 illustrates a structured approach to fog architecture [33], above or beyond the physical and transport

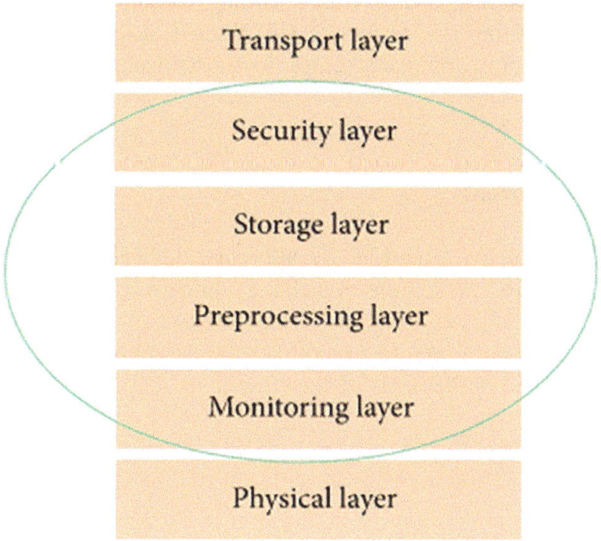

Fig. 2 An IoT portals with fog architecture

layers. Power, reactions, and calculations are all under the watchful eye. Sensor data is filtered, stored, and evaluated in the preprocessing layer. Data transfer, delivery, and recovery are both available via the temporary storage layer. Finally, the authentication layer protects data integrity and anonymity by performing encryption and decryption. Before submitting data to the cloud, data is monitored and preprocessed at the network's edge.

"Edge computing" and "fog computing" are also interchanged. The later term goes back much farther and is broader. Fog computing, which was created by Cisco, refers to sensors and access points in edge devices This machine learning model focuses on mechanical components, motors, pumps, and lamps. The aim is to reduce the data to as much as feasible at the network edge. The computer architecture of Fig. 2 is identical to the architectural diagram; as a result, we don't have edge computing.

Finally, there is a fuzzy distinction between protocols and devices. We often use protocols and programs. For both cloud and fog architectures, we will use the traditional stack architecture (architectural diagram seen in Fig. 2).

2.3 Social IoT

Let's take a look at a different paradigm: social IoT (SIoT). In this article, we interpret social interactions between artifacts in the same way as humans do [14]. The three major aspects of a SIoT scheme are as follows: (i) The SIoT can be navigated. We can begin with one device and work our way through all of the devices attached to it. Using a social network of IoT devices, it is simple to explore different devices and utilities. (ii) There is a need for trustworthiness (relationship strength) between devices (similar to friends on Facebook). (iii) We will research the social networks of IoT devices utilizing templates close to those used to study human social networks.

2.3.1 Basic Components

In a traditional social IoT situation, we approach devices and services like bots, enabling them to shape connections and change them over time. This will enable us to seamlessly allow the machines to work together to complete a complex mission.

We'll need a number of integrated modules to make this model work. Let's have a peek at some of the key elements of such a scheme.

Object Recognition

we need a special form of object identification. Traditional values such as the MAC ID, an IPv6 ID, or any other system-specific type can be used to define the object.

Metadata

In addition to an ID, we need certain metadata about the system that defines its type and function. This is necessary in order to create acceptable relationships with the unit and to properly position it in the IoT universe.

Security Controls

This is identical to Facebook's "friend list" settings. The types of devices that may attach to a computer might be limited by the device's owner. Owner monitors are the most common term for these.

Domain Discovery

A framework like this is similar to a service cloud, in which we need dedicated folders to store specifics of devices that have certain types of services. It becomes important to maintain these directories current so that devices can learn about each other.

Partnership Management

This partnership manager is responsible for tracking other equipment. Additionally, it maintains track of the gadgets it is authorized to connect with. It makes sense for an example.

Service Composition

It advances the society paradigm. It's a device built to maximize co-occurrence, e.g., if an individual puts a power sensor on her air conditioner, it will provide everyday energy usage statistics. If you were to integrate such use patterns into the social model, much more useful information might be discovered. Users would be recommended, for example, to use the most resources in their local community or on Facebook.

2.3.2 Architecture that is Representative

Any of the systems have a cloud structure the server is part of it, working as the main point of contact. Three standard server designs the database which has data regarding each computer's properties, as well as relationships. Computer communications, condition monitoring, and system operation were done by the second layer.

The network serves consumers. Directly over the object layer enables computers to communicate and share data. Collecting facts from social media the consumer is manipulated by the system

3 In the Internet of Things, Privacy Is Critical

3.1 Motivation

IoT privacy is a critical protection problem that requires complete consideration from academic and business researchers. Proposing guidelines and control systems to address privacy in IoT is important. IoT is also used in a number of technologies, including virtual patient control, electricity use management, traffic management, and smart parking systems. Users need security of personal details relating to their travel, behaviors, and relationships with other users in both of these applications.

3.2 Challenges

Any approach or mechanism for privacy in IoT must resolve the following challenges.

3.2.1 Tracking and Profiling

Associating a name with a specific person poses a risk since it may contribute to identification and monitoring. As a consequence, one of the most important tasks is to ban such operation in the IoT and introduce any preventative steps.

3.2.2 Tracking and Localization

Another threat is localization, which occurs as devices attempt to evaluate and record a person's position over time and space. Designing protocols for communications with IoT that prohibit such behavior is one of the main challenges of protection solutions for IoT. In e-commerce applications, profiling details about a specific person to infer preferences by connection with other profiles and data is quite general. The balance of commercial priorities in targeting and data collection with consumer protection criteria is a huge task [34, 35].

3.2.3 Data Transfer that is Reliable

Another layer of encryption is to guarantee the data is shared safely through a public network without revealing personal details, avoiding unwanted data collection regarding items and individuals.

3.3 Discussion and Existing Approach

This section covers current efforts to ensure IoT privacy, particularly with body sensor networks [36]. The research focusing on network and data protection for the Internet of things focuses on identity safety, location privacy, and node manipulation and adding/removing assaults, while current research on privacy-enhancing technology looks at forward and backward vulnerabilities as well. Various tactics and places of change to fix those [37].

In [38], the writers conducted a review of current IoT implementations. The authors suggested a translation of their modules into a standard device model while also defining and analyzing differentiating behavioral trends of sensor data produced in this article. Almost all applications collect position and time details, according to the report. The data collected may be in a range of forms, including video and audio. The writers undertook a study of existing privacy protections. In addition, possible risks to consumer privacy in participatory sensing as a consequence of unregulated dissemination of personal details to untrustworthy individuals were addressed. The authors also connected their findings to a proposed common interface model for assessing security in mobile participatory sensing applications.

[39] address protection risks and safety in IoT architectures in detail. The debate starts with a thorough glance at the IoT's layered design. At each stage of the design, privacy and protection risks are carefully investigated, and the latest state of the art in IoT architecture research is closely investigated. Privacy, eavesdropping, and man-in-the-middle (middle) assault issues differ depending on the purposes. In addition, the writers look at the emerging EU IoT regulations. It's critical to comprehend the IoT architecture's control realms. At all levels of architecture, a person must be able to monitor his or her knowledge, according to EU legislation. Further analysis is required to explain how this method of regulation is theoretically assisted. More research into the energy dimensions of privacy and threats is needed.

The developers of [40] looked at privacy improvements in IoT through a number of application domains. Future smart home security specifications are addressed in depth. In addition, the developers proposed an acceptable protection framework for IoT. The gateway design has been selected as the best choice for devices with minimal resources and high system availability. This design will operate vital smart home functions by applying sophisticated control algorithms on a relatively efficient processor. Other architectures that are scrutinized for IoT include middleware architecture and cloud architecture, in addition to gateway architecture. One creativity is focused on gateways: technical maintenance to aid reliability and another on long-maintaining software and firmware updates to ensure ongoing support.

[41] outline efforts to keep user privacy in the IoT by properly labelling IoT details. Privacy properties are assigned to the sensed data, allowing trustworthy control access dependent on sensitivity. Tagging is very costly due to the resource requirements of IoT, and this paper addresses the issues with tagging resource-constrained IoT. Four properties of the physical interaction, functional details, and sensor tags are key requirements for effective privacy protection of IoT. Apart from these four, the introduction of IFC data tags renders connected processes more straightforward. Security issues are discussed thoroughly in Manet's mobile topology; shared wireless medium, significant resource shortage, and dynamic network topology are all barriers to security. And with these obstacles, multilayer defense still works. Security concerns and cutting-edge technology multihop transmission strategy on cell nodes an approach that includes complete protection, detection, and response cover the three layers [42].

It talks about enabling technology like RFID as well as privacy considerations, such as RFID, too. RFID technology is thought to be useful for object monitoring and inventory management. To extend this to people, rules and regulations must be in place, as well as stringent imposition to guarantee acceptance, as it may be manipulated. According to the writers, concerns with technical and social issues must be overcome in order to use RFID to allow IoT [43].

4 What Is Federated Learning, and How Does It Work?

The flow of federated learning is depicted in the diagram below.

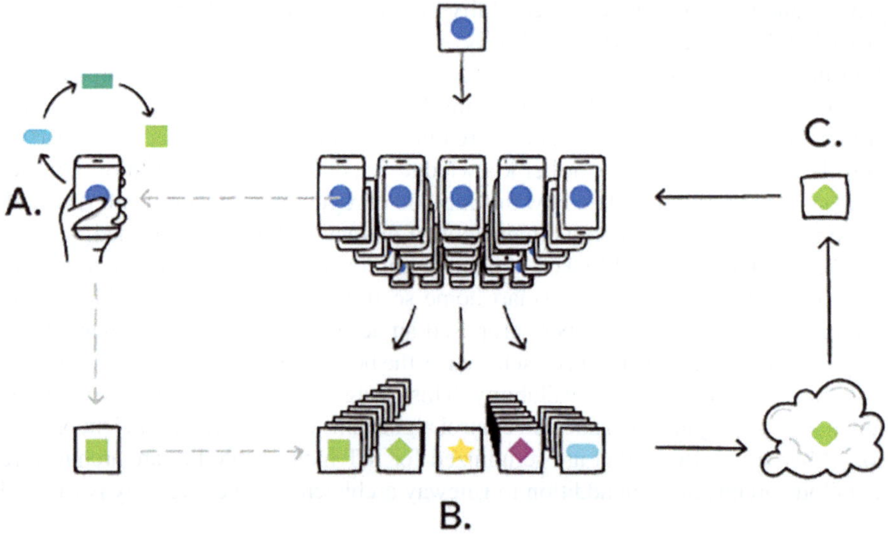

Fig. 3 Depending on your use, your phone customizes the model locally. (**a**) The changes of several users are combined (**b**) to form a consensus adjustment (**c**) to the shared model, and then the process is repeated. (Figure by *Google AI Blog*)

Based on your use and info, your phone customizes the model locally (A). The changes of other users that use that platform are then merged (B) to form a consensus modification (C) to the shared model, and the process is repeated. FL complies with GDPR in that the central server does not exchange or archive the user's data or privacy from each computer; however, it only shares and stores what has been learned from the devices' updates, which is then combined into an updated model and sent back to the client. The mechanism then repeats itself.

Federated learning is based on the "FedAvg" Federated averaging algorithm [44]. FedAvg is Google's [44] original vanilla Federated learning algorithm for solving federated learning problems. Since then, several variants of FedAvg algorithms have been developed to solve many of the federated learning problems in [45], such as "FedProx," "FedMa," "FedOpt," "Scaffold," and so on.

The following is a high-level summary of how the FedAvg algorithm operates.

The aim of each round of FedAvg is to reduce the global model's objective w, which is simply the amount of the weighted average of the local system failure.

(i) A random subset of clients/devices is sampled.
(ii) Each client receives the server's global model.
(iii) Clients run Stochastic Gradient Descent (SGD) on their own loss functions in parallel and submit the resulting model to the server for aggregation.
(iv) The server then uses the sum of these local models to change the global model.
(v) The procedure is then replicated for n more rounds of contact.

4.1 *Federated Learning's Obstacles*

With such a beneficial invention comes a slew of issues that must be resolved [45, 46]. The following four points curated from [46] are some of the most critical challenges that I feel need to be tackled.

4.1.1 Noise Versus Precision Trade-Off

Using Differential Privacy (DP), we will apply noise to data to improve privacy security. However, with DP, we had to compromise the model's results. As a consequence, a trade-off must be made in order to introduce the necessary amount of noise while not jeopardizing the model's output.

4.1.2 System and Statistical Heterogeneity

Since training on heterogeneous systems is difficult, it's critical to ensure that federated learning scales well over all devices, regardless of platform form. The word "dissimilarity of statistical knowledge" applies to a device's failure to extract a global statistical trend, resulting in populations, measurements, or outcomes that vary from those of other devices.

4.1.3 Communication Bottlenecks

The connectivity expense of getting the models to the device should be moderately minimal, since this could have an effect on the FL climate, where one device may become disabled due to communication bottlenecks, causing the federated training phase to slow. Dropping stragglers (devices that failed to compute training within the prescribed time window) and model compression/quantization to reduce bandwidth costs are two approaches to solving connectivity bottlenecks.

4.1.4 Poisoning

There are two types of poisoning:

1. Data poisoning: During a federated training phase, many clients can contribute by contributing their on-device training data, and it's challenging to detect/prevent malicious clients from sending malicious/fake data to poison the training method and therefore poison the model.
2. Poisoning models: In comparison to data poisoning, malicious clients tamper with the obtained model's gradients/parameters before submitting it back to the central server for aggregation, resulting in the global model being heavily polluted with invalid gradients throughout the aggregation phase.

4.1.5 Trade-Off Between Reliability and Safety

Using Secure Multi-Computation (SMPC) and Differential Privacy (DP) improves the privacy security capabilities of federated learning, but this protection comes at an expense. Clients must encrypt model parameters until submitting them back to the central server using SMPC; as a result, additional computing resources are needed for encryption, compromising the model's training performance. Since DP adds noise into the model and results, some precision is lost. As a result, seeking a good trade-off between SMPC and DP is still a work in progress in federated learning.

4.2 *Problems Solved by Federated Learning Have the Following Characteristics*

The optimal problems for federated learning to tackle, according to Google study, have three characteristics.

4.2.1 First Characteristic

Training on real-world data from mobile devices has a huge benefit over training on proxy data available in the data center. As a server creates a single machine learning model, it incorporates data from several users to generate a single generalized model. Since consumers utilize mobile devices in a number of ways, the model should be generic enough to accommodate this heterogeneity. Unfortunately, a standardized model would not improve the user experience; however, a personalized model that seems to have been designed exclusively for the device will. This degree of personalization is possible, thanks to federated learning, which provides the idea that the device was designed exclusively for the consumer.

4.2.2 The Second Property

That this data is either vulnerable to privacy or huge in scale (in comparison to the size of the model), so logging it to the data center only for the purpose of model training is not recommended (in service of the focused collection principle). It's not realistic to require the customer to send vast volumes of data to the server in order to construct a generic model. The customer would incur extra expenses as a result of this. Also, particularly for apps that need personal user details, the user is likely to refuse uploading private data to help create a model. In contrast to centralized learning, federated learning is a strong choice in situations where data is private or broad in size. With this increase in complexity, federated learning faces a new obstacle. The services accessible to mobile devices are minimal. Working with vast volumes of data takes time and, as a result, needs more electricity. TensorFlow Federated, a miniature version of TensorFlow ideal for on-device testing, was created to address this problem. Only while the system is idle and wired in and has a free wireless link can training take effect.

4.2.3 Third Property

Labels on the data can be inferred naturally from user activity with controlled activities. Here are two problems that match the previous three properties, according to Google research:

4.2.3.1 Image classification to determine which images will be seen several times or exchanged in the future.
4.2.3.2 Language models that can enhance speech recognition and text entry on touchscreen keyboards by optimizing decoding, next-word prediction, and also whole-response prediction.
4.2.3.3 A model is trained using labeled data in supervised learning such that the model recognizes the labels of all training samples. The labels of the user's private data must be accessible for federated learning to function with supervised learning.

4.3 Federated Machine Learning Meets Intellectual IoT

The potential implementations and tasks of federated machine learning will be discussed in this section.

4.3.1 Intelligent IoT Services Utilizing Federated Machine Learning

The use of machine learning technology is vital for intelligent utilities, such as self-driving cars and the use of drones, as well as healthcare. In traditional cloud-based machine learning applications, the data must be shared to a centralized cloud that is used to train artificial intelligence (AI) models. Many academics and industry people have been concerned about processing power and data privacy issues because of attempts to move the AI from the data center to the IoT edge. Internet of things (IoT) is a modern way to train a global machine learning model with distributed computation capabilities that still keeps the data on the devices.

In Fig. 3, a federated machine learning architecture is used to power intelligent Internet of things applications (base station and internet of things device nodes). To aggregate an IoT machine, each device manages and updates a local model. The global and local models are also updated over time until it comes to a consensus.

4.3.2 During Each Learning Round, the Following Three Phases Are Performed in Order

Platform Selection

The machine only selects a subset of IoT devices for locality model training and maturation (e.g., those with good channel conditions and important software updates).

Local Model Upload

Every system will update the local models based on the most recently accessed global algorithm (e.g., stochastic gradient) and then upload the created global updates through wireless uplink transmission.

Global Model Download

The server upgrades the local versions and broadcasts the downlink transmissions to the chosen IoT computers, allowing them to do the next round of learning. Training is not feasible for the amount of data collected by a single IoT unit or method. Resolving the issue of data autonomous islands, the IoT devices' variety of data, lets a huge quantity of privacy-sensitive data aid trainable data into a global model for IoT networks.

4.3.3 The Following Are Some of the Benefits of Federated Deep Learning in Intelligent IoT Networks

User Privacy and Data Protection Are Preserved

The obtained data is not communicated with the IoT or with any IoT products. In addition, federated machine learning guarantees device and data secrecy.

Collective Training

Federated machine learning will combine computing capital and gathered data through a wide range of dispersed IoT devices to create a global machine learning model collaboratively. As a consequence, each system obtains a machine learning model that is superior to what it can do on its own.

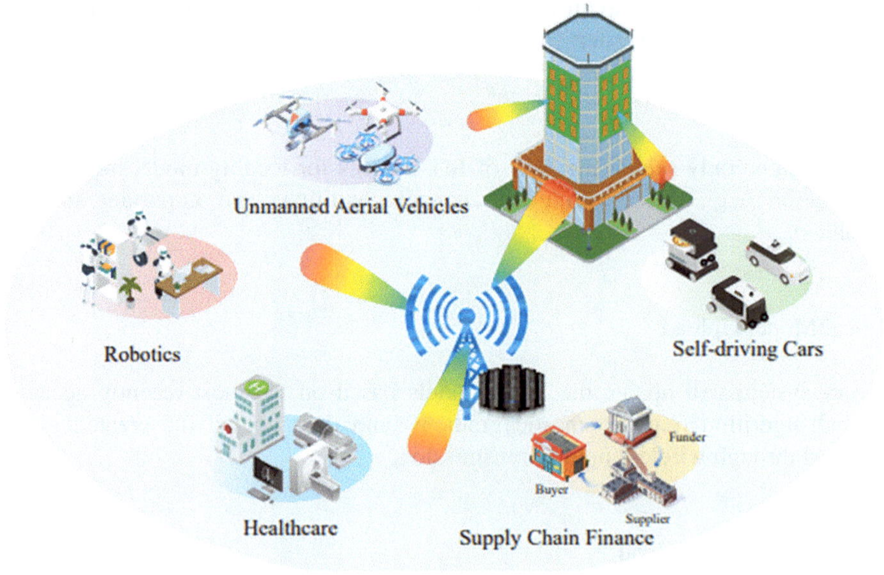

Fig. 4 Reconfigurable artificial surface for intelligent machine learning

Reduced Network Latency

Since federated machine learning doesn't need IoT devices to transmit heavy amounts of data to a remote cloud, it reduces network latency drastically [47].

These benefits render federated machine learning a promising approach for distilling knowledge directly at the network edge, allowing machine learning to reach its maximum potential in a variety of thrilling intelligent IoT applications.

4.3.4 Federated Machine Learning Enables Key Intelligent IoT Applications

There are various other upcoming applications, such as self-driving cars, UAVs, and healthcare and robotics, in which federated machine learning (FML) is supposed to greatly enhance their decision-making method, as stated in Fig. 4.

The following are several descriptions of standard implementations in various industries:

Self-Driving Cars

One of the key elements behind the self-driven vehicle market is the Internet of things (IoT). Each sensor has the capacity to generate a hundred gigabytes of data per second (such as video cameras, lidar, and radar). Innovative machine learning tasks, including adaptive obstacle avoidance and environment control more intelligent vehicles must have a higher degree of latency when processing dynamic situations [48]. Furthermore, data protection is important for self-driving vehicles because vehicle data usually includes a tone of personal details regarding consumers. Federated machine learning holds promise in terms of effectively and safely teaching AI models through smart vehicles, as well as reducing network latency by using information at the network edge [49].

Unmanned Aerial Vehicles (UAVs)

Unmanned Aerial Vehicles (UAVs) have found a host of civilian and industrial uses in recent years, including traffic control, freight transport, and virtual reality (VR). To help more intelligent IoT implementations, UAVs are supposed to be equipped with pervasive wireless networking and embedded knowledge in future communication systems [50]. Since continuous contact with the remote cloud center is needed when performing AI tasks in UAV networks, there can be substantial communication and signaling delays. Federated machine learning, by relocating the AI-infused engines to the network edges, shows how effective it is in decreasing network latency

Healthcare

Utilizing real-world healthcare evidence, machine learning has shown tremendous promise in enhancing healthcare efficiency, such as designing diagnostic methods and forecasting disease threats. Patients' healthcare data is highly vulnerable and confidential, and it is shielded by stringent regulatory policies across the world [51]. Concentrated machine learning shows its use in collaboratively training healthcare AI models while protecting the integrity of patient data. To further safeguard the delivery, the local updates, use homomorphism encryption a federated deep learning-based healthcare network that safeguards one patient's information. In the midst of the Covid pandemic, the Ford Institute has built an in-home virus monitor powered by federated machine learning.

Robotics

The Internet of Robotic Systems (IoTS), which combines robotics and IoT, assists intelligent robots with vision and movement and runs at lightning level. IoRT is versatile due to its ability to interactively adapt to events across many manufacturing industries. Single robots, given their limitations, depend on local data and have significant reaction time lags to unusual circumstances. Capacity gains of federated ML enable for progressively complex and complicated distributed processes to be undertaken with increasingly nuanced results.

Supply Chain Finance

When combined with machine learning and IoT, supply chain finance will accelerate capital and knowledge movements across the supply chain, closing the financial distance between customers and suppliers. Industrial monitoring data in supply chain networks, in particular, must be transferred from various business organizations and can be used to detect and extract underlying trends using cutting-edge machine learning technology. The supply chain data, which is highly critical and therefore essential to business organizations, is susceptible to security problems in such a distributed environment. Federated machine learning offers a modern way to prevent knowledge disclosure while still lowering supply chain financing efficiency and credit risk [52].

4.3.5 Federated Machine Learning's Coordination Problem

While federated machine learning has a lot of potential for intelligent IoT, it still has a lot of drawbacks, such as limited battery capacity, unbalanced data, heterogeneous computing capacities, and so on. Due to scarce radio infrastructure and iterative propagation of high-dimensional model-update parameters, connectivity bottleneck becomes a key obstacle in federated machine learning for intelligent IoT applications. In particular, the federated machine learning mechanism usually involves hundreds of contact rounds, through which multiple machines upgrade local models through a wireless multi-access channel. Making use of a standard orthogonal multi-access system, interaction model leads to exponentially the contact loads on the IoT units a large volume of traffic congestion will result in excessive network latency [53–56]. We will bring forward a new coordinated machine learning radio access scheme based on AirComp to minimize the necessary computation and communication in channels. The reconfigurable intelligent surface also improves AirComp's range by mitigating unfavorable Wi-Fi factors.

4.4 Artificial Intelligence (AI) as a Defense Mechanism and IoT Security Requires

According to Girma, the massive and expansive use of IoT devices brings a fresh way of looking at computing. Enabled machines are increasingly possible, and the risk of needing a more secure cyber infrastructure to manage and address both data at rest and data in motion is important. Additionally, IoT data gathering, transferring, and encoding all performed at a very high degree of confidentiality as well [57, 58]. Where did she/he go? The application of several AI frameworks to the extremely flexible cyber-physical infrastructure and several linked devices in a network-rich setting has proven crucial for developing a viable IDS and IPS. Many companies have implemented artificial intelligence (AI) as part of their vulnerability posture in order to improve their overall cybersecurity. The key purpose of using standard protection methods has been rule-based. Currently, AI is helping businesses deal with a relentless threat from the cyber. Given the nature of data collecting and computing both occur in the cloud, the use of AI as a part of a cyber-policy is becoming a default, and leading to this at a high stage. One of the big advantages of AI/ML is real-time vulnerability monitoring and checking of security risks [50, 59].

Advantages of Demonstration though artificial intelligence (AI) and machine learning (ML) promise to be helpful to both science and cybersecurity, they present completely opposite characteristics. They are operating 24/7 to implement an evolving AI/malware that incorporates modern methods and exploits new openings. They also hack malware that they create, which they will use to hack their own networks, rendering their malware more successful; the more time and data input, the greater the ensuing harm. Other limitations include the high expense, the complexity of rendering it human-like, the reality that it must be changed, the lack of originality, and reacting to machine assaults. to a certain extent, it is based on its dataset coming from various sources the data must be correct to benefit from if you have none of these items It's becoming more prevalent, which may allow malicious insiders or external adversaries to target and poison the data, but can also results that can be far more exacting and hard to detect.

4.5 Federated Machine Learning of Over-the-Air Computing

This segment introduces AirComp, a novel multiple access scheme for quick model aggregation in federated machine learning that combines communication and computation.

4.5.1 Over-the-Air Computation Rules

The standard solution, calculating values at just one machine-mode units, usually results in significant transmission overheads. Theoretically speaking, AirComp turned out to be a good way to aggregate wireless data by using the superposition of signals to understand it over the airwaves; one can refer to nomography. If you have nomograms for the raw numbers, calculate the amount of all post-processed values to AirComp; the user-supplied input data is loaded in until the nomographic functions are applied. to display the complete number of the original encoded data at all IoT devices. You can preprocess all of the data to obtain the function. The AirComp connectivity and processing method for sensing was originally conceptualized for practical computing over multi-access networks. It shows that practical values can be obtained instead of cancelled out by intervention. The current decade has seen extensive and broad research into AirComp. Information theory has investigated various distributions of the input results, including total of the squares, normalized sum of squares, skewed sum of squares, etc. Other designs also investigated ways to reduce the effect of beamforming on aggregation error. Furthermore, channel details have been taken into account for AirComp. to the introduction of AirComp have been thoroughly.

5 Conclusions

To our distributed computing post, we talked about the capacity of federated machine learning to solve the crucial issues of IoT, to maintain data protection, to alleviate congestion, and to take maximum advantage of the power of distributed capital IoT expand on this, that, in addition, AirComp and RIS are two new developments that we have found that can help with the restricted connectivity bandwidth problem in machine learning model federation.

References

1. I. Mohd Zaki, H. Rosilah, The implementation of internet of things using test bed in the UKMnet environment. Asia Pac. J. Inf. Technol. Multimed. **8**, 1–17 (2019)
2. R. Hassan, S.S. Nori, N.E. Othman, The improvement of the protection for 6LoWPAN in IoT through non-causal hash function scheme, in *Proceedings of the 2018 15th International Conference on Electrical Engineering/Electronics, Computer, Telecommunications and Information Technology (ECTI-CON)*, Chiang Rai, Thailand, 18–21 July 2018, pp. 600–603
3. A. Whitmore, A. Agarwal, L. Da Xu, The Internet of Things—A survey of topics and trends. Inf. Syst. Front. **17**, 261–274 (2015)
4. Z.M. Ali, M.A.B.M. Arshad, M.A. Bakar, POLIoT: Internet Of things framework in managing network threats at metro polytechnic tasek gelugor, in *Proceedings of the 2018 Cyber Resilience Conference (CRC)*, Putrajaya, Malaysia, 13–15 November 2018, pp. 1–4
5. P. Jain, F. Adrangi, M. Venkatachalam, *Cellular IoT Network Architecture*. Google Patents US 10,623,942 B2, 14 April 2020
6. S. Korade, V. Kotak, A. Durafe, A review paper on internet of things (IoT) and its applications. Int. Res. J. Eng. Technol. **6**, 1623–1630 (2019)
7. F. Wu, T. Wu, M.R. Yuce, An internet-of-things (IoT) network system for connected safety and health monitoring applications. Sensors **19**, 21 (2019)
8. S.R. Pungus, J. Yahaya, A. Deraman, N.H.B. Bakar, A data modeling conceptual framework for ubiquitous computing based on context awareness. Int. J. Electr. Comput. Eng. **9**, 5495–5501 (2019)
9. M.H. Alsharif, R. Nordin, N.F. Abdullah, A.H. Kelechi, How to make key 5G wireless technologies environmental friendly: A review. Trans. Emerg. Telecommun. Technol. **29**, e3254 (2018)
10. H. Zhang, X. Lu, Vehicle communication network in intelligent transportation system based on internet of things. Comput. Commun. **160**, 799–806 (2020)
11. I.S. Udoh, G. Kotonya, Developing IoT applications: Challenges and frameworks. IET Cyber Phys. Syst. Theory Appl. **3**, 65–72 (2018)
12. B. Afzal, M. Umair, G.A. Shah, E. Ahmed, Enabling IoT platforms for social IoT applications: Vision, feature mapping, and challenges. Future Gener. Comput. Syst. **92**, 718–731 (2019)

13. K.B. Letaief, W. Chen, Y. Shi, J. Zhang, Y.-J.A. Zhang, The roadmap to 6G–AI empowered wireless networks. IEEE Commun. Mag. **57**, 84–90 (2019)
14. H. Li, K. Ota, M. Dong, Learning IoT in edge: Deep learning for the Internet of things with edge computing. IEEE Netw. **32**, 96–101 (2018)
15. Q. Yang, Y. Liu, T. Chen, Y. Tong, Federated machine learning: Concept and applications. ACM Trans. Intell. Syst. Technol. **10**, 1–19 (2019)
16. X. Wang, Y. Han, C. Wang, Q. Zhao, X. Chen, M. Chen, In-edge AI: Intelligent zing mobile edge computing, caching and communication by federated learning. IEEE Netw. **33**, 156–165 (2019)
17. J. Kang, Z. Xiong, D. Niyato, S. Xie, J. Zhang, Incentive mechanism for reliable federated learning: A joint optimization approach to combining reputation and contract theory. IEEE Internet Things J. **6**, 10700–10714 (2019)
18. K. Yang, T. Jiang, Y. Shi, Z. Ding, Federated learning via over- the-air computation. IEEE Trans. Wirel. Commun. **19**(3), 2022–2035 (2020)
19. G. Zhu, Y. Wang, K. Huang, Broadband analog aggregation for low-latency federated edge learning. IEEE Trans. Wirel. Commun. **19**, 491–506 (2020)
20. M.M. Amiri, D. Gunduz, *Federated Learning Over Wireless Fading Channels* (IEEE Trans. Wireless Commun, 2020). https://doi.org/10.1109/TWC.2020.2974748
21. Q. Wu, R. Zhang, Towards smart and reconfigurable environment: Intelligent reflecting surface aided wireless network. IEEE Commun. Mag. **58**(1), 106–112 (2020)
22. C. Huang, A. Zappone, G.C. Alexandropoulos, M. Debbah, C. Yuen, Reconfigurable intelligent surfaces for energy efficiency in wireless communication. IEEE Trans. Wirel. Commun. **18**, 4157–4170 (2019)
23. S. Han, Y. Huang, W. Meng, C. Li, N. Xu, D. Chen, Optimal power allocation for SCMA downlink systems based on maximum capacity. IEEE Trans. Commun. **67**, 1480–1489 (2019)
24. I. Mashal, O. Alsaryrah, T.-Y. Chung, C.-Z. Yang, W.-H. Kuo, D.P. Agrawal, Choices for interaction with things on internet and underlying issues. Ad Hoc Netw. **28**, 68–90 (2015)
25. O. Said, M. Masud, Towards internet of things: Survey and future vision. Int. J. Comput. Netw. **5**(1), 1–17 (2013)
26. M. Wu, T.-J. Lu, F.-Y. Ling, J. Sun, and H.-Y. Du, Research on the architecture of internet of things, in *Proceedings of the 3rd International Conference on Advanced Computer Theory and Engineering (ICACTE' 10)*, vol. 5 (IEEE, Chengdu, China, 2010), pp. V5-484–V5-487
27. R. Khan, S.U. Khan, R. Zaheer, S. Khan, Future internet: the internet of things architecture, possible applications and key challenges, in *Proceedings of the 10th International Conference on Frontiers of Information Technology (FIT' 12)* (2012), pp. 257–260
28. H. Ning, Z. Wang, Future internet of things architecture: Like mankind neural system or social organization framework? IEEE Commun. Lett. **15**(4), 461–463 (2011)
29. M. Weyrich, C. Ebert, Reference architectures for the internet of things. IEEE Softw. **33**(1), 112–116 (2016)
30. J. Gubbi, R. Buyya, S. Marusic, M. Palaniswami, Internet of Things (IoT): A vision, architectural elements, and future directions. Futur. Gener. Comput. Syst. **29**(7), 1645–1660 (2013)
31. F. Bonomi, R. Milito, P. Natarajan, J. Zhu, Fog computing: A platform for internet of things and analytics, in *Big Data and Internet of Things: A Road Map for Smart Environments*, (Springer, Berlin, 2014), pp. 169–186
32. F. Bonomi, R. Milito, J. Zhu, S. Addepalli, Fog computing and its role in the internet of things, in *Proceedings of the 1st ACM MCC Workshop on Mobile Cloud Computing* (2012), pp. 13–16
33. I. Stojmenovic, S. Wen, The fog computing paradigm: scenarios and security issues, in *Proceedings of the Federated Conference on Computer Science and Information Systems (FedCSIS' 14)* (IEEE, Warsaw, Poland, 2014), pp. 1–8
34. M. Aazam, E.-N. Huh, Fog computing and smart gateway based communication for cloud of things, in *Proceedings of the 2nd IEEE International Conference on Future Internet of Things and Cloud (FiCloud' 14)*, Barcelona, Spain, August 2014, pp. 464–470

35. L. Atzori, A. Iera, G. Morabito, SIoT: Giving a social structure to the internet of things. IEEE Commun. Lett. **15**(11), 1193–1195 (2011)
36. J. Zhou, Z. Cao, X. Dong, A.V. Vasilakos, Security and privacy for cloud-based IoT: Challenges. IEEE Commun. Mag. **55**(1), 26–33 (2017)
37. N. Aleisa, K. Renaud, Privacy of the internet of things: A systematic literature review, in *Proceedings of 50th Hawaii International Conference on System Sciences*, Waikoloa, HI, USA, 2017
38. D. Christin, A. Reinhardt, S.S. Kanhere, M. Hollick, A survey on privacy in mobile participatory sensing applications. J. Syst. Softw. **84**(11), 1928–1946 (2011)
39. J. Veijalainen, D. Kozlov, Y. Ali, Security and privacy threats in IoT architectures, in *Proceedings of 7th International Conference on Body Area Networks*, Oslo, Norway, September 2012
40. H. Lin, N. Bergmann, IoT privacy and security challenges for smart home environments. Information **7**(3), 44 (2016)
41. D. Evans, D.M. Eyers, Efficient data tagging for managing privacy in the internet of things, in *Proceedings of 2012 IEEE International Conference on Green Computing and Communications*, Besancon, France, November 2012, pp. 244–248
42. H. Yang, H. Luo, Y. Fan, S. Lu, L. Zhang, Security in mobile ad hoc networks: Challenges and solutions. IEEE Wirel. Commun. **11**(1), 38–47 (2004)
43. B. Khoo, RFID as an enabler of the internet of things: issues of security and privacy, in *Proceedings of 4th International Conference on Cyber, Physical and Social Computing*, Dalian, China, 2011, pp. 709–712
44. *Communication-Efficient Learning of Deep Networks from Decentralized Data*
45. *Advances and Open Problems in Federated Learning*
46. *A survey on security and privacy of federated learning*
47. K. Kishor, P. Nand, P. Agarwal, Notice of retraction design adaptive subnetting hybrid gateway MANET protocol on the basis of dynamic TTL value adjustment. Aptikom J. Comput. Sci. Inform. Technol. **3**(2), 59–65 (2018)
48. K. Kishor, P. Nand, Review performance analysis and challenges wireless MANET routing protocols. Int. J. Sci. Eng. Technol. Res. **2**(10), 1854–185 (2013). ISSN 2278 – 7798
49. D. Tyagi, D. Sharma, R. Singh, K. Kishor, Real time 'driver drowsiness' & monitoring & detection techniques, Int. J. Innov. Technol. Explor. Eng. **9**(8) (2020). ISSN 2278-3075
50. S.P. Yadav, Emotion recognition model based on facial expressions. Multimed. Tools Appl. (2021). https://doi.org/10.1007/s11042-021-10962-5
51. K. Kishor, S. Saxena, S. Yadav, S. Yadav, Study and development of air monitoring and purification system, Vivechan Int. J. Res. **10**(2), (2019). ISSN 0976-8211
52. A. Jain, K. Kishor, Financial supervision and management system using Ml algorithm. Solid State Technol. **63**(6), 18974–18982 (2020)
53. A. Girma, Analysis of security vulnerability and analytics of internet of things (IOT) platform. Inf. Technol. New Gener. **738**, 101–104 (2018)
54. S.P. Yadav, Vision-Based detection, tracking and classification of vehicles. IEIE Trans. Smart Process. Comput. SCOPUS **9**(6), 427–434 (2020). https://doi.org/10.5573/IEIESPC.2020.9.6.427. ISSN: 2287-5255
55. K. Kishor, P. Nand, P. Agarwal, Subnet based ad hoc network algorithm reducing energy consumption MANET. Int. J. Appl. Eng. Res. **12**(22), 11796–11802 (2017). ISSN 0973-4562
56. E.P. Yadav, E.A. Mittal, D.H. Yadav, IoT: Challenges and issues in Indian perspective, in *3rd International Conference On Internet of Things: Smart Innovation and Usages (IoT-SIU)*, Bhimtal, India, 2018
57. S.P. Yadav, S. Yadav, Image fusion using hybrid methods in multimodality medical images. Med. Biol. Eng. Comput. **58**, 669–687 (2020). https://doi.org/10.1007/s11517-020-02136-6
58. T.V. Phan, M. Park, Efficient distributed denial-of service attack defense in SDN-based cloud, in *IEEE*, 2019
59. H.S. Anderson, A. Kharkar, B. Filar, Evading machine learning malware detection. Int. J. Appl. Eng. Res. (2017). Black Hat, pp. 1–6

Federated Learning for an IoT Application

Deena Nath Gupta, Rajendra Kumar, and Shamsul Haque Ansari

1 Introduction

Internet of Things (IoT) is the new computing environment for the next generation of applications. The applications developed in the past few years look mainly for scalability. One of the inherent properties of scalability is heterogeneity. The applications, if they want to be scalable, should take care of heterogeneity issues. For example, different thermal sensors may produce temperature in different forms, i.e., 35 degrees or 35.0 degrees or 0035.00 degrees. The mechanism should be capable of storing all the information in a global common form. This needs some intelligent modification in the entries of received tables [1].

Artificial intelligence (AI) was introduced with the core concept of taking smart and intelligent decisions on its own. Many industrial applications are benefited with the automation provided by AI [2]. Algorithms are written in such a way that the applications can figure out the next move according to the current situation. The AI algorithms were complex in nature and hence they are not suitable for the constrained devices used in the IoT applications [3]. An IoT application mainly works under a constrained environment, i.e., low battery, low power, and low storage [4] In an IoT environment, the number of participating nodes are very large. In order to get a specific command work in an IoT environment, every node needs to be fitted with the same algorithm. The large number of deployments of complex algorithms are costly affairs for a constrained environment [5]. Hence, AI is somehow not suitable here.

Since every single node in an IoT environment is smart enough to make decisions, a learning approach is required at the initial phase of the development of the

D. N. Gupta (✉) · R. Kumar · S. H. Ansari
Jamia Millia Islamia, New Delhi, India
e-mail: rkumar1@jmi.ac.in

S. P. Yadav et al. (eds.), *Federated Learning for IoT Applications*, EAI/Springer Innovations in Communication and Computing,
https://doi.org/10.1007/978-3-030-85559-8_4

concept of IoT. It is well known that there is no human intervention in the functioning of an IoT system, also called machine-to-machine communication. Since machines are the building blocks of an IoT environment, they should be capable of making themselves smart again and again according to the need of applications over time [6]. The concept of machine learning (ML) works in this context. ML is a practice that includes taking inputs from each individual machine then after developing a common update, providing this update to every individual machine [7]. This is also known as a centralized learning method. The only drawback of this approach is its centralized nature.

Meanwhile, a new concept, known as federated learning (FL), emerged to make the learning suitable for the highly distributed nature of applications in an IoT environment [8]. In FL, an application can be modified as per user's requirement at device level only. There will be no requirement for data sharing at global level for getting an update because each device uses their own data to compute the update. This will help to protect user's data privacy as well. The module update in a federated setting is twofold. The first phase consists of sharing a mechanism to each potential device at a regular interval. Every device uses their own data to perform computations locally. The results from this local update are then shared to the global platform. In the second phase, the global platform combines the results received from each device to get a global update model. This globally accepted model is then sent to every device for update [9].

Since IoT comprises a large number of constrained devices, the arrangement in FL seems to be suitable for IoT applications. In an IoT environment, the devices are used for some specific purposes only. Hence, it would be more convenient to inject them with update modules. The learning methods may be enabled in the devices right after installation. This will help them to reduce the required communication with central authority. The devices can themselves produce a periodic update and send this update to the central authority for cumulative action needed. With this, the IoT environment benefits both the ways, less communication and high security of one's personal data [10–13].

The rest of the chapter is structured as follows. The details about federated learning along with edge learning and decentralized learning are provided in Sect. 2. Section 3 provides the challenges in the way of implementation of federated learning in an IoT application. Threats and possible countermeasures are provided in Sect. 4. The authors concluded their work in Sect. 5.

2 Federated Learning

Data are the new assets for today's individual and hence to protect them is the need of hour. The safest place for one's data is inside their device. The moment any data is sent out from the device, it becomes vulnerable to attack. Although many algorithms are already available that claim the security of data on the wire, yet one cannot avoid the vulnerability issue. Seeing the current pace of development, one also cannot ignore being globally available for updates. This situation raises to an obvious question - “is there any mechanism available that can guarantee the data protection along with providing all global updates?”

Yes, federated learning is the answer. Unlike machine learning, federated learning provides the opportunity of distributed learning to the participating devices. Currently federated learning is in its conceptual phase. Many applications are switching their learning method to FL from ML [14]. Suppose there are “N” number of applications that use a global model for their functioning. Every application has their own data generated from many events. Applications are not supposed to share these data with other applications. Suppose they are having D1, D2, D3, D4, ….., Dn data respectively. Each application will use their respective data to train the module. These module updates will then be aggregated at global level to generate a globally acceptable module update [15].

Unlike earlier centralized learning methods, FL only performs the aggregation Magr on the received individual update modules. If the module updates from the ith device are denoted as Mi and the parameters are supposed to have only mathematical inputs, then FL can aggregate the inputs received from all the applications as:

$$\mathrm{Magr} = \left(\sum_{i=0}^{n} \mathrm{Mi} \right) / n$$

This will give the average value for all the parameters used inside a module. The global module will then use this aggregation to update the module. Update module in FL may differ with the update module in ML.

Also, there are two different architectures presented in a federated setting. One is called vertical federated learning and another is called horizontal federated learning. Individual applications share their gradient estimate to central authority individually in a horizontal setting, while in vertical setting the gradient estimate of one application is encrypted with other applications in a serial fashion to make a single cumulative update. Note that no data are shown to either of the devices in any circumstance [16]. The encryption over gradient estimates is shown in Fig. 1.

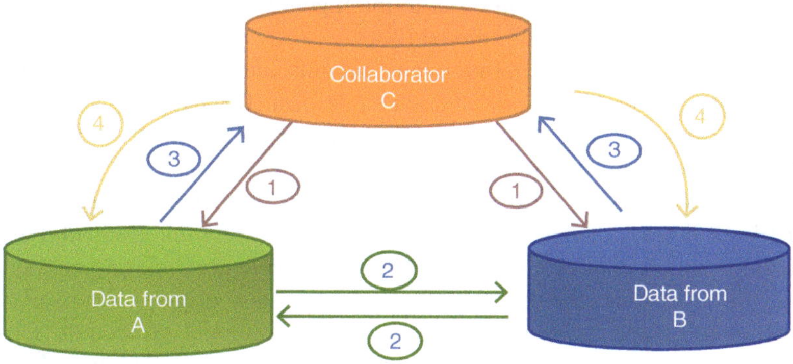

Fig. 1 Secure transfer of gradient estimate

The process may be subdivided into four parts. (1) Public keys of participating applications are broadcasted. (2) Intermediate results are exchanged among participating applications using the asymmetric key cryptography. (3) Cumulative gradient descent is computed and forwarded to central authority. (4) The final updated module is sent back to every participating application. The process of secure transfer of gradient estimate is demonstrated in Fig. 1.

There are two main questions the FL is facing. The first question is – "is the effect of errors in data collection be reflected in the updates?" The decentralized learning is performed on different locations and from all locations data may not be of the same form. A small variation in precision making may lead to an avalanche effect. The second question is – "what is the security of one's local update when they transmit it over the network?" Previously, data were sent to the wire. Many cryptographic solutions are available for data protection [17]. In FL, the applications are going to share their individual model update. These models should also need to transmit securely over the wire.

2.1 Edge Learning in an IoT Environment

Federated learning for an IoT application is similar to the edge computing in a wireless network. The sensors deployed in different geographical areas can have data at very frequent intervals. They may produce undesired or useless data maximum of the time. For example, a temperature sensor produces almost similar temperatures for a certain period of time. Hence, there should be a program running in all the edge devices for averaging the inputs of a particular time frame. The devices at the edge layer in an IoT environment are intelligent enough to take decisions on their own, whereas in earlier cloud computing environments, only the central authority takes the decision. This decision shift is known as federated learning [18, 19].

Multiaccess edge computing (MEC) enables the applications with the ability to make inferences and predictions. These predictions are useful in formulating the state update for end users. Also, it helps the actuators to take the right action. Although MEC is very useful in order to intelligently reduce the required communication, it does not fit in the FL model [20].

FL never allows for data sharing in any circumstance. In this scenario, the edge devices need a distributed stochastic gradient descent (DSGD) approach to train the applications locally. This approach benefits the IoT environment in both the way, less transmission required as well as data protection.

Communication-efficient distributed ML is the need of the hour. Figure 2 depicts a scenario of heterogeneous device communication, where the gradient descent of sensed information may be either sent from a mobile device, a desktop, or a laptop. Quantization, Sparsification, and local updates are among the well-known approaches to reduce the required communication and data protection. The local gradient estimate needs to be communicated at the global level [21]. In an IoT application, the number of local learning might be generated every hour by lakhs of

Fig. 2 Heterogeneous devices communication

devices. Quantization reduces the number of sharing for every outcome from learning of each iteration. Sparsification advocates sending the best update only. Choosing the most effective update among all the available updates over a period is required here. One can also choose to reduce the frequency of communication to parameters only while going for local update [22].

2.2 *Learning for Decentralized IoT Applications*

Coordination and cooperation are what is required for heterogeneously distributed devices in a scalable IoT application [23]. One of the possible solutions includes agent-based learning. Many agents are there to learn from the edge of a wireless network and then by including the inputs from every agent, a team decision is taken to send to the global platform. Although it is very crucial in FL to reach a team decision, yet it is not that easy. This is an open research problem – "To achieve multi-device decision-making, is there any possibility to derive a robust algorithm with arbitrary inputs and having uncertainties across agents?" [24].

Many researchers studied this field and presented their work. A well-known article "Team decision problems" was presented by Radner in 1962 that solved a linear objective problem. Again, in 1972, Marschak and Radner presented "Economic theory of teams" to solve some simple cases of that time. These could be seen as references to design the team update problem in FL. Consider a multiagent scenario having "N" number of agents. Let us assume that, at some particular point of time if an agent "i" takes a decision "Di" using an information "Ii", then with Si (an arbitrary function from the information space to the decision space), the decision taken can be represented as:

$$\mathrm{Di} = \mathrm{Si}(\mathrm{Ii})$$

The information available at the edge of an IoT application is expected to be an incomplete estimation of the true representation. This leads to the imperfection of available information for an agent. A cooperative scenario always tries to maximize the objective function from every agent in order to make an absolute decision.

A quantization scheme, known as "separate digital," is used to reduce the required bandwidth to the allowed channel limit. In this scheme, the computing agents convert their gradient estimates into bits and then communicate independently to channel codes. Although it takes less bandwidth, the transmission rate is high. The parameter server needs only the average of the gradient estimate. Hence, an analog signal, without coding, can be transmitted in synchronization with the wireless channel. The parameter server will itself perform averaging of the received and desired values [25, 26].

2.3 *Federated Learning for an IoT Application*

FL is best suited for an IoT application till date due to its efficiency for updates and data protection model. Some of the IoT applications are worth noting here such as wearable devices, autonomous vehicles, and smart homes. These applications produce a lot of data at each iteration and also need real time up to date information for their smooth operation. The core challenges include expensive communication, systems heterogeneity, statistical heterogeneity, and privacy concerns. Limited bandwidth availability of an IoT application restricts the devices to transmit data at regular intervals. If an application generates a large amount of data all the time, then it could be expensive in terms of IoT. FL converts this large data into a small chunk of information by performing averaging of the values [8, 9, 27]. Averaging of the values in a federated setting is depicted in Fig. 3.

Different clusters in an IoT application may behave differently. For example, network connectivity in different applications might be 3G, 4G, 5G, or Wi-Fi. This leads to the variation in transmitting signal strength. Also, other hardware constraints may differentiate the capability of participating devices. Power supply to different clusters also varies. There may be a possibility of drop out arising in a network during a transmission. Hence, the mechanism used in FL should be fault tolerant. Sometimes it may also happen that the gradient estimates received are very less in number. This happens due to drop out in remote areas. Hence, the global updated module based on less participation may not be suitable for everyone. The FL should call for the next update immediately if it gets less response than a threshold [28, 29].

An IoT application generates data in a highly non identically distributed manner. The data generated from an IoT application may violate (i.i.d.) assumption, independent and identically distributed assumption, of FL. Also, the model update at a regular interval may be a threat to the data protection strategy of FL. The secure multiparty computation (SMC) might be useful here. If this will be the scenario, applicability of SMC in an IoT application will need to be taken care of. Differential

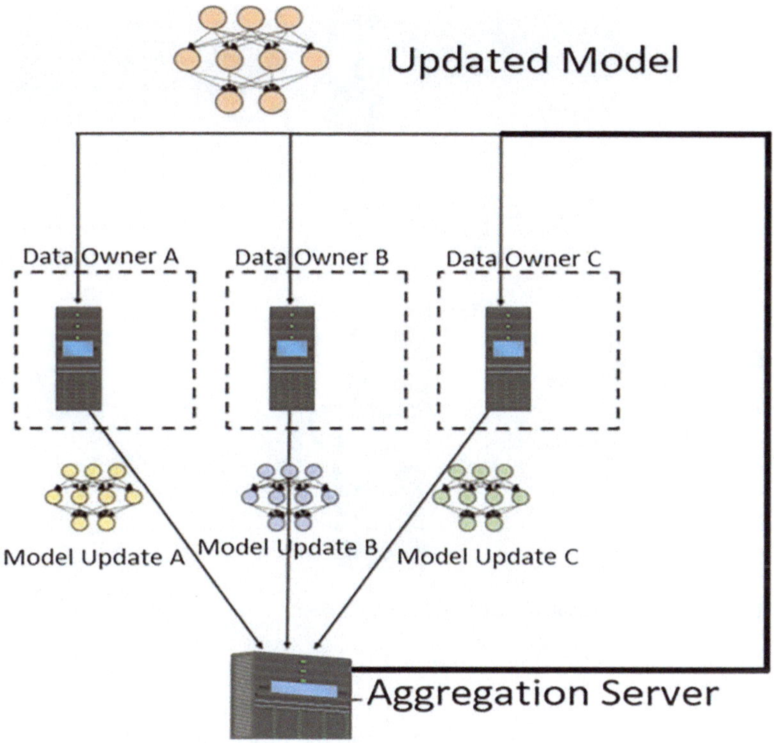

Fig. 3 Averaging of the values in a federated setting

privacy is another possible approach, mathematically proven, the only drawback is that the performance of the system will go down because of complexity [14, 15].

3 Challenges in FL

Many applications leak the data of their users for financial gain. Exchange of user's data is also practiced by some applications mutually. The law enforcement agencies have already made many laws against this malpractice. Although many of the companies have been punished for private data leakage, data sharing among companies is not stopped completely. Laws to impose strict controls on data collection and transactions have been made in many countries. China's Cyber Security Law and the General Provisions of Civil Law will be implemented in China in 2017. In 2018, the European Union (EU) adopted General Data Protection Regulation (GDPR). California Consumers Privacy Act (CCPA) was enacted in the state of California, U.S. recently [30–34].

In an IoT environment, where a huge amount of data are generated continuously, safe storage is the main issue. The only place where the data are most secure is

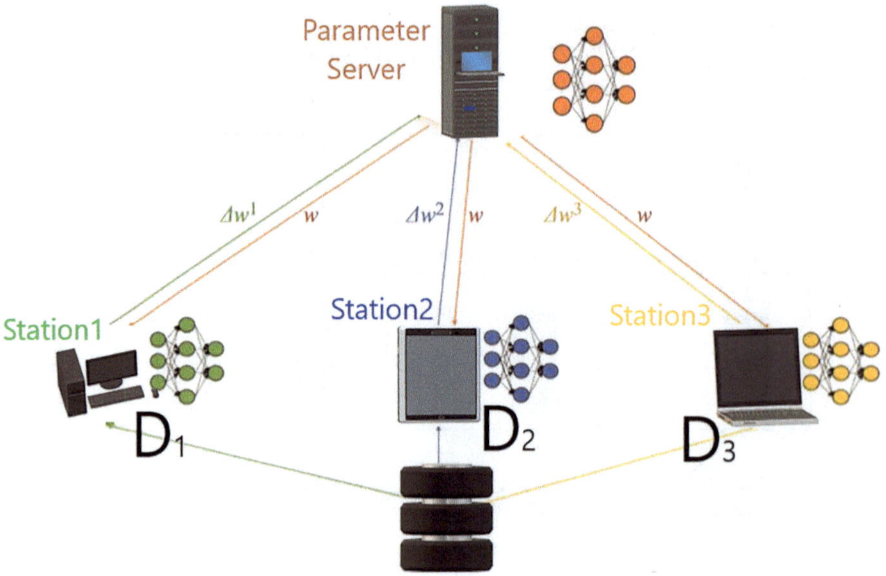

Fig. 4 Illustration of a distributed machine learning system

within the application itself. This gives rise to the processing of data at the edge of the application only. FL also advocates almost the same methodology. Hence the suitability of FL in an IoT environment can be considered higher than any other learning mechanism. Some challenges yet exist in the implementation of FL in an IoT environment. The research community of computer science already presented many variations in the field of automated learning. Some of them are, privacy-preserving ML, privacy-preserving DL, collaborative ML, collaborative DL, distributive ML, distributive DL, federated optimization, and privacy preserving data analytics [20, 35–38]. Distributed machine learning system is depicted in Fig. 4.

3.1 Inability to Inspect Training Data

The global update in FL is not based on the received data from the participating application. So, FL cannot choose the parameters to train the model by itself. The parameter selection for training a model will remain local. This may lead to different parameter selection by different applications. At the time, when the global update cumulates all the local updates, this will cause a big problem. Modeling a deep neural network (DNN) is not suitable in this scenario.

3.2 *Participation of Big Players*

The data generated by every participating application are not the same in size. One application may generate thousands of data while another may generate billions of data. Similarly, data points are also varying for different applications. The trained model by using a large dataset will produce a more accurate model than by using the small dataset. This leads to an unfair situation while distributing the global model update to all the participating applications in equal proportion. Fairly, the big players should get more access or the early access to the global update.

3.3 *Cheating Behavior*

Honesty is the best policy. However, the honesty of every participating application cannot be guaranteed. Unlawful activities might be performed by one or more participating applications to get more access to the global update. Some of the bad practices are producing dummy data and reporting false training results.

3.4 *Communication Failure*

In vertical federated learning (VFL), nodes are connected in a serial manner. The updates of the next application depend on the updates from the previous application. Unlike in HFL, any communication loss in one application can affect all further updates. The architecture of vertical federated learning is depicted in Fig. 5.

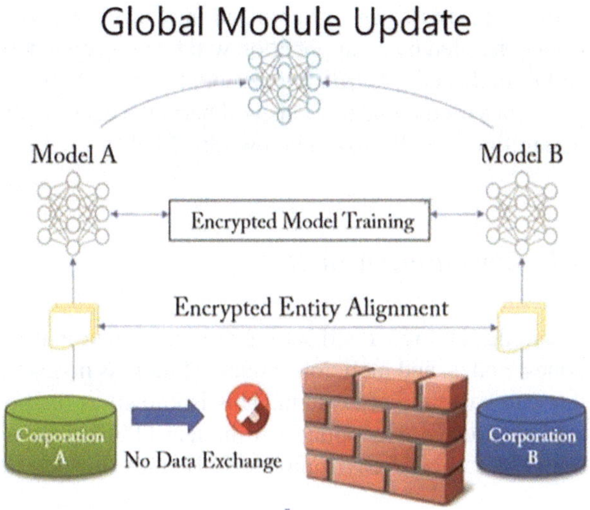

Fig. 5 Vertical federated learning architecture

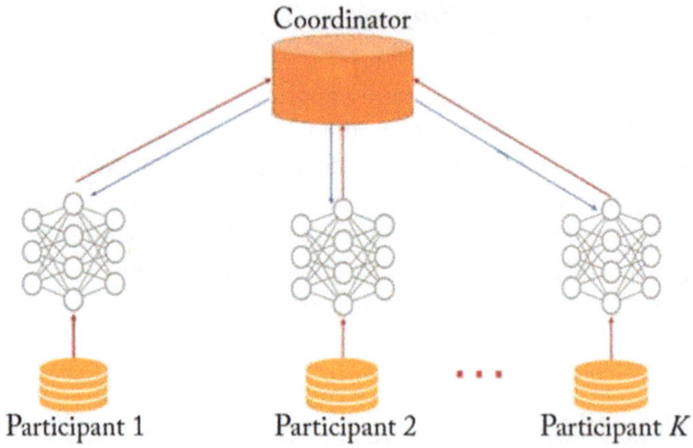

Fig. 6 Horizontal federated learning architecture

3.5 Low Participation

In horizontal federated learning (HFL), nodes are communicated in parallel manner. At some time, connections of multiple nodes might be broken. These nodes will not be able to send their gradient estimate to the global platform. The global modal update will be unfair for all the participating applications in this case. The architecture of horizontal federated learning is depicted in Fig. 6.

4 Threats and Possible Countermeasure

Unlike ML, in FL, an IoT application need not worry about the security of its data. Data generated by an application will be in its own periphery only. The only vulnerability in this case is fraudulent modal update. A malicious application may inject a poisonous modal update. This will harm the ecosystem in several ways. Some of the possibilities are discussed below [35, 39–44].

4.1 Poisoning Attack

The nodes in a federated setting are segregated in three forms, honest nodes, semi-honest nodes, and malicious nodes. Malicious nodes train their model in such a way that can harm the ecosystem. This is known as the poisoning attack in a federated setting. Trusted execution environment (TEE) is seen as one of the possible solutions. In a trusted environment, only authenticated actors can participate. Before sending any model update, it will be necessary for the application to get themselves registered into the global update team.

4.2 *Backdoor Attack*

Model replacement is a technique used by some of the malicious players in an environment to replace the received model from the previous player to some other ill-fitted model. The methodology used in this type of update is blockchain where all the players are connected in a linked list fashion. The global update is the resultant of the cumulative action taken on different local updates. The attack in this arrangement is known as backdoor attack. Some mechanisms used to guard the learning environment are defensive distillation and adversarial training regularization.

4.3 *Equation-Solving Attack*

The malicious node trains its model with a particular equation. After the global update, the equation can be solved by the malicious node to gain some useful information about all the trained models or their data. This is known as the equation-solving attack in a federated setting. Suppose there is a node X residing somewhere in the chained environment, when the model update request reaches this node, it injects its own modified set of requests. All the nodes after this node X will assume this modified set of requests as originally sent by the controlling authority and they update this model cumulatively. In return, when the global model reaches this node X, it can respond to the requests (injected earlier by itself) to gain the access. Theoretically, for an N-dimensional linear model, an adversary can steal it with (N + 1) queries. Such a problem can be formalized as solving € from (x, h€(x)). To defend against the equation-solving attack, a secure two-party computation approach for vertical federated linear regression and classification is proposed by introducing the concept of k-secure.

4.4 *Extreme Poisoning Attack*

In order to compute the average of a group of gradients, a coordinator is required. In a secure system, this coordinator is assumed not to see the individual gradients of the participants. In some cases, the coordinator may inject a malicious model by itself and circulate the update to all the participants. This can even harm the whole working environment. This is called an extreme poisoning attack. Bonawitz et al. [14] propose "trimmed mean," where the gradients are trimmed coordinate-wisely to prevent extreme poisoning.

5 Conclusion

The authors discussed a distributed learning opportunity for resource constrained and highly distributed IoT applications. The earlier technique for learning was centralized. The centralized learning is not suitable for an IoT application. There are two main issues, first is data protection, and second is heterogeneity. To protect the user's data, the concept of federated learning emerges. Federated learning is applied on the participating devices only. Hence, the devices need not to share their data anywhere. Also, two types of federated learning have been discussed, parallel and serial. In horizontal federated learning, the devices send the model updates in parallel while in vertical federated learning, the devices perform the model update by using the updated models of their counterparts. At last, the authors discussed the challenges and threats in the way to successfully implement federated learning in an IoT application. Countermeasures to these threats are theoretical and need to be formulated in future.

References

1. D.N. Gupta, R. Kumar, Lightweight cryptography : An IoT perspective. Int. J. Innov. Technol. Explor. Eng **8**(8), 700–706 (2019)., [Online]. Available: https://www.ijitee.org/download/volume-8-issue-8/
2. S.G. Shaila, A. Vadivel, V. Naksha, Introduction to the World of Artificial Intelligence (2019), pp. 359–379. https://doi.org/10.4018/978-1-5225-9902-9.ch019
3. B. Hoanca, K.J. Mock, "rtificial intelligence-based cybercrime. Encycl. Crim. Act. Deep Web, 36–51 (2020). https://doi.org/10.4018/978-1-5225-9715-5.ch003
4. M. Katagi, S. Moriai, Lightweight cryptography for the Internet of Things. Sony Corp. 7–10 (2008), [Online]. Available: http://www.iab.org/wp-content/IAB-uploads/2011/03/Kaftan.pdf
5. J. Hernandez-Castro, G. Avoine, Cryptanalysis of ubiquitous computing systems, *Proc. 18th Mediterr. Electrotech. Conf. Intell. Effic. Technol. Serv. Citizen, MELECON* 2016, no. April (2016), pp. 18–20. https://doi.org/10.1109/MELCON.2016.7495307
6. D.N. Gupta, R. Kumar, A. Kumar, Efficient encryption techniques for data transmission through the internet of things devices, in *IoT and Cloud Computing Advancements in Vehicular Ad-Hoc Networks*, ed. by V. Jain, O. Kaiwartya, N. Singh, R. S. Rao, 1st edn., (IGI Global, Pennsylvania, 2020), pp. 203–228
7. V. Mkrttchian, S. Kanarev, L.A. Gamidullaeva, Machine learning and cyber security, Encycl. Crim. Act. Deep Web, 1034–1042 (2020). https://doi.org/10.4018/978-1-5225-9715-5.ch070
8. Q. Yang et al., Federated Learning
9. Q. Yang, Y. Liu, T. Chen, Y. Tong, Federated Machine Learning: Concept and Applications (Feb. 2019), [Online]. Available: http://arxiv.org/abs/1902.04885
10. R. Khan, S. U. Khan, R. Zaheer, S. Khan, Future internet: The internet of things architecture, possible applications and key challenges, *Proc. - 10th Int. Conf. Front. Inf. Technol. FIT* (2012), pp. 257–260. https://doi.org/10.1109/FIT.2012.53
11. K. Fan, Y. Ren, Z. Yan, on Blockchain, *2018 IEEE Int. Conf. Internet Things IEEE Green Comput. Commun. IEEE Cyber, Phys. Soc. Comput. IEEE Smart Data* (2018), pp. 1349–1354. https://doi.org/10.1109/Cybermatics
12. D.N. Gupta, R. Kumar, Generating random binary bit sequences for secure communications between constraint devices under the IOT environment, *INCET* (2020), pp. 1–6

13. M.D. Green, Cryptography for secure and private databases: Enabling practical data access without compromising privacy, *ProQuest Diss. Theses* (2009), p. 170.
14. K. Bonawitz et al., Towards Federated Learning at Scale: System Design (Feb. 2019), [Online]. Available: http://arxiv.org/abs/1902.01046
15. M.A. Thomas, D.S. Abraham, D. Liu, An Assessment of Federated Machine Learning for Translational Research (2020), pp. 123–142. https://doi.org/10.4018/978-1-7998-1879-3.ch006
16. J. Konečný, H.B. McMahan, F.X. Yu, P. Richtárik, A.T. Suresh, D. Bacon, Federated Learning: Strategies for Improving Communication Efficiency (Oct. 2016), [Online]. Available: http://arxiv.org/abs/1610.05492
17. D.N. Gupta, Sponge based Lightweight Cryptographic Hash Functions for IoT Applications (2021), pp. 1–5
18. S. Tuli et al., HealthFog: An ensemble deep learning based Smart Healthcare System for Automatic Diagnosis of Heart Diseases in integrated IoT and fog computing environments. Futur. Gener. Comput. Syst. **104**, 187–200 (2020). https://doi.org/10.1016/j.future.2019.10.043
19. D.N. Gupta, R. Kumar, Networking in IoT: Technologies used, security threats and possible countermeasures. Int. J. Sensors, Wirel. Commun. Control. **10** (Jan. 2021). https://doi.org/10.2174/2210327910666210106090826
20. A. Jolfaei, K. Kant, H. Shafei, Secure data streaming to untrusted road side units in intelligent transportation system, *Proc. - 2019 18th IEEE Int. Conf. Trust. Secur. Priv. Comput. Commun. IEEE Int. Conf. Big Data Sci. Eng. Trust* (2019), pp. 793–798. https://doi.org/10.1109/TrustCom/BigDataSE.2019.00115
21. M.R. Arul, V. Sathiyamoorthi, Introduction to machine learning and its implementation techniques, (Samuel 1959), 334–358 (2019). https://doi.org/10.4018/978-1-5225-9902-9.ch018
22. L.M. Candanedo, V. Feldheim, D. Deramaix, Data driven prediction models of energy use of appliances in a low-energy house. Energ. Buildings **140**, 81–97 (2017). https://doi.org/10.1016/j.enbuild.2017.01.083
23. M.S.H. Ansari, M. Mehrotra, Development of smart cities and its sustainability: A smart city framework. Int. J. Innov. Technol. Explor. Eng **8**(11), 646–655 (2019). https://doi.org/10.35940/ijitee.K1703.0881119
24. N. Mishra, L.P. Verma, P.K. Srivastava, A. Gupta, An analysis of IoT congestion control policies. Procedia Comput. Sci. **132**, 444–450 (2018). https://doi.org/10.1016/j.procs.2018.05.158
25. P. Rakshit, I. Nath, S. Pal, Application of IoT in Healthcare, in *Principles of Internet of Things (IoT) Ecosystem: Insight Paradigm*, ed. by S.-L. Peng, S. Pal, L. Huang, (Springer International Publishing, Cham, 2020), pp. 263–277
26. J. Kaur, A. Kumar, M. Bansal, Lightweight cipher algorithms for smart cards security: A survey and open challenges, *4th IEEE Int. Conf. Signal Process. Comput. Control. ISPCC* 2017, vol. 2017-Janua (2017), pp. 541–546, https://doi.org/10.1109/ISPCC.2017.8269738
27. D.N. Gupta, R. Kumar, Distributed key generation for secure communications between different actors in service oriented highly dense VANET, in *Cloud and IoT-Based Vehicular Ad Hoc Networks*, John Wiley & Sons, Ltd (2021), pp. 221–232
28. H.B. McMahan, E. Moore, D. Ramage, S. Hampson, B. A. y Arcas, Communication-Efficient Learning of Deep Networks from Decentralized Data (Feb. 2016), [Online]. Available: http://arxiv.org/abs/1602.05629
29. H. Brendan McMahan, E. Moore, D. Ramage, S. Hampson, B. AgüeraAg, A. Arcas, Communication-Efficient Learning of Deep Networks from Decentralized Data (2017)
30. O.S.L. Tan, R.G. Vergara, R.C.W. Phan, S. Khan, N. Khan, Cybersecurity laws in Malaysia, Encycl. Crim. Act. Deep Web, 435–448 (2020). https://doi.org/10.4018/978-1-5225-9715-5.ch030
31. O.S.L. Tan, R.G. Vergara, R.C.W. Phan, S. Khan, N. Khan, Human trafficking and cyber Laws in Malaysia, Encycl. Crim. Act. Deep Web. 518–532 (2020). https://doi.org/10.4018/978-1-5225-9715-5.ch035
32. T.U. Rehman, International context of cybercrime and cyber law, Encycl. Crim. Act. Deep Web, 412–423 (2020). https://doi.org/10.4018/978-1-5225-9715-5.ch028

33. S.P. Yadav, K.K. Agrawal, B.S. Bhati, F. Al-Turjman, L. Mostarda, Blockchain-based cryptocurrency regulation: An overview. Comput. Econ. 0123456789 (2020). https://doi.org/10.1007/s10614-020-10050-0
34. T. Wang, Y.-Y. Wang, An Overview of Recent Development in Privacy Regulations and Future Research Opportunities (2020), pp. 88–101. https://doi.org/10.4018/978-1-7998-5728-0.ch005
35. K.K. Sreelakshmi, A. Bhatia, A. Agrawal, Securing IoT Applications Using Blockchain (2020), pp. 56–83. https://doi.org/10.4018/978-1-7998-2414-5.ch004
36. J. Zhang, Y. Zhao, J. Wu, B. Chen, LPDA-EC: A lightweight privacy-preserving data aggregation scheme for edge computing, *Proc. - 15th IEEE Int. Conf. Mob. Ad Hoc Sens. Syst. MASS* (2018), pp. 98–106. https://doi.org/10.1109/MASS.2018.00024
37. D. Asok, P. Chitra, B. Muthurajan, Privacy Preserving Machine Learning and Deep Learning Techniques (2019), pp. 222–235. https://doi.org/10.4018/978-1-5225-9902-9.ch012
38. R. Boussada, M.E. Elhdhili, L.A. Saidane, A Lightweight Privacy-Preserving Solution for IoT: The Case of E-Health, *Proc. - 20th Int. Conf. High Perform. Comput. Commun. 16th Int. Conf. Smart City 4th Int. Conf. Data Sci. Syst. HPCC/SmartCity/DSS 2018* (2019), pp. 555–562. https://doi.org/10.1109/HPCC/SmartCity/DSS.2018.00104
39. V. Rajasekaran, J. Mohana, Commerce Computer Science SECURITY ISSUES WITH POSSIBLE SOLUTIONS Assistant Professor, Department of Computer Science, PSG College of Arts And Science, Coimbatore Assistant Professor, Department of Computer Science, PSG College of Arts And Science, (1), 48–50 (2015)
40. A. Langiu, C.A. Boano, M. Schub, K. Romer, UpKit: An open-source, portable, and lightweight update framework for constrained IoT devices, *Proc. - Int. Conf. Distrib. Comput. Syst.*, vol. 2019-July, no. iii (2019), pp. 2101–2112. https://doi.org/10.1109/ICDCS.2019.00207
41. A. Da Veiga, Achieving a Security Culture (2019), pp. 72–100. https://doi.org/10.4018/978-1-5225-7847-5.ch005
42. R.G. Engoulou, M. Bellaïche, S. Pierre, A. Quintero, VANET security surveys. Comput. Commun. **44**, 1–13 (2014). https://doi.org/10.1016/j.comcom.2014.02.020
43. S. Rao, A.K. Verma, T. Bhatia, Evolving Cyber Threats, Combating Techniques, and Open Issues in Online Social Networks (2020), pp. 219–235. https://doi.org/10.4018/978-1-7998-5728-0.ch012
44. M.S.H. Ansari, M. Mehrotra, Securing M2M communication in Smart Cities, *2020 International Conference for Emerging Technology (INCET)* (2020), p. 4

Some Observations on the Behaviour of Federated Learning

Vishal Kaushal and Sangeeta Sharma

1 Introduction

Advances in power of machine learning have brought with them significant concerns about data privacy. In particular, this is true when it comes to training machine learning models with information obtained from user interaction with devices such as smartphones. Uploading data to a server and using that to train models is the traditional process for training a machine learning model. This training method works just fine as long as there is no issue with data privacy. This method becomes unsuitable when it comes to training machine learning models where personally identifiable information is involved.

Training models on a centralized server often means that you need a huge volume of storage space, as well as world-class protection to prevent security issues. But imagine if you could use local data stored at the user end for training the model. Price and delay still affect traditional machine learning techniques. It is difficult to solve these issues, and data is enormously distributed. Concerning computation and communication, this raises problems with machine learning techniques.

V. Kaushal (✉) · S. Sharma
Department of Computer Science and Engineering, National Institute of Technology Hamirpur, Hamirpur, Himachal Pradesh, India
e-mail: Vishal_phdcse@nith.ac.in; sangeetas@nith.ac.in

S. P. Yadav et al. (eds.), *Federated Learning for IoT Applications*, EAI/Springer Innovations in Communication and Computing,
https://doi.org/10.1007/978-3-030-85559-8_5

2 Federated Learning

FL also called decentralized learning is a technique that downloads the current model and learning the local data on the device. These locally trained updated models are sent to the central server, using encrypted communication, where a large pool of devices share their model and then improve the shared model by averaging with the other devices [1].

We can apply FL to any edge nodes, and it can vastly improve various critical domains such as finance, healthcare, smart home transportation, etc. Recently, some medical students and researchers in different parts of the world worked together to develop a working machine that diagnoses COVID-19 from chest scan [2]. Another application would be autonomous driving vehicle training and urban route planning in the field of transport network. Similarly, for smart home applications, edge devices in different homes can learn collaboratively about context-conscious policies using an FL framework [3].

At the user level in FL, the downloaded data trained by learning to improve the model [4]. Aggregation of updated models is done at the server end. Each user sends its updated model to the server. A consolidated model is then transferred back to the devices [5]. Low cost is the benefit for the use of federated machine learning model on devices such as cellphones and sensors [6]. Figure 1 shows the architecture of FL.

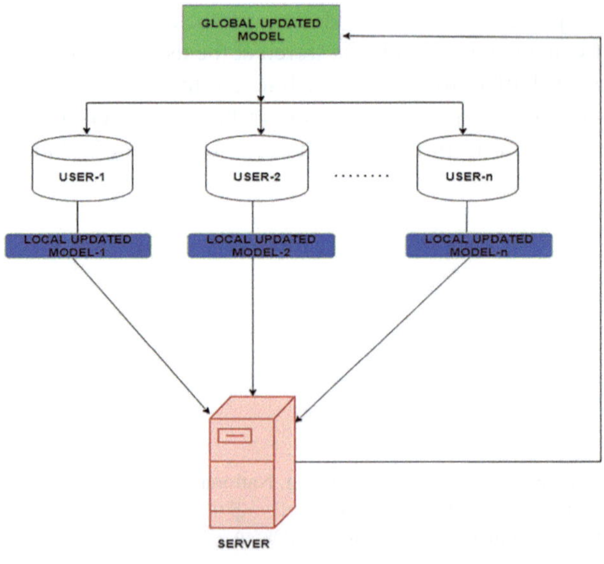

Fig. 1 Architecture of federated learning

3 Different Architectures of FL

The categorization of FL architecture is done based on *data distribution*.

3.1 Sample-Based Federated Learning (SBFL)

SBFL is also known as horizontal federated learning where datasets have similar feature space but different in terms of data or sample. For example, in the case of Google keyboard, mobile users use the same features but have different training data. Another example is where Google presents an SBFL approach for handling android cellphone updates. Consider two banks which have very different types of customer from their corresponding area. The common set of the customer is small, but business is, however, quite the same. So overall, the feature space is the same, but there is a difference in data or sample. As shown in Fig. 2, the following four steps take place:

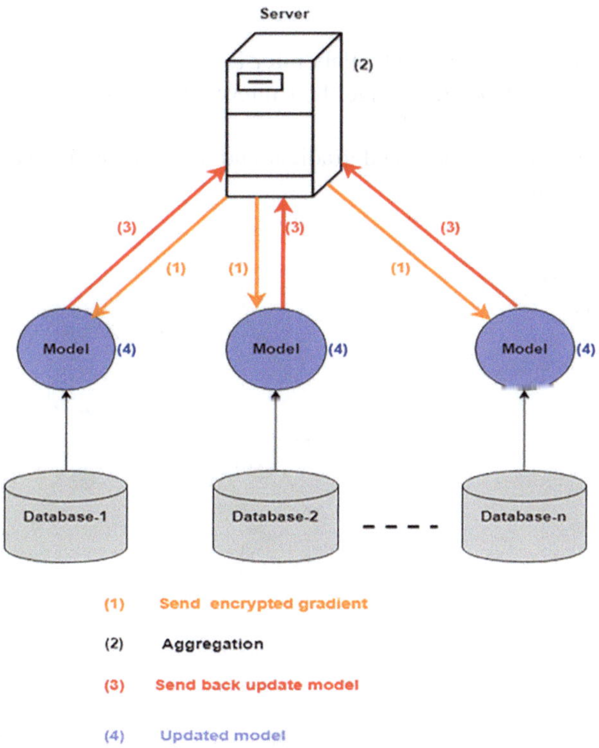

Fig. 2 Sample-based federated learning architecture

1. User locally computes training gradients and after encryption sends the result to the server.
2. Aggregation takes place at the server's end without knowing the user's information.
3. Server computes the aggregated result and sends it back to the users.
4. With the help of decrypt gradients, users update their models.

3.2 Feature-Based Federated Learning (FBFL)

Feature-based federated learning (FBFL), sometimes known as vertical federated learning, is the reverse of sample-based federated learning where datasets have the same data or sample space but vary in terms of feature space. Let us consider an example of two companies in a city that have data of the same group of people but both the companies use different feature spaces. As shown in Fig. 3, P and Q are two companies which provide training to a machine learning model together; both companies have their data, but they cannot directly exchange data due to privacy and security reasons. For confidentiality of data collaborator, R takes part throughout the training process.

1. The Public key is sent by the Collaborator R to P and Q.
2. P and Q encrypt and interchange their intermediate results for loss calculation and gradients.
3. P and Q compute the encrypted gradient and loss and send the encrypted value to the collaborator

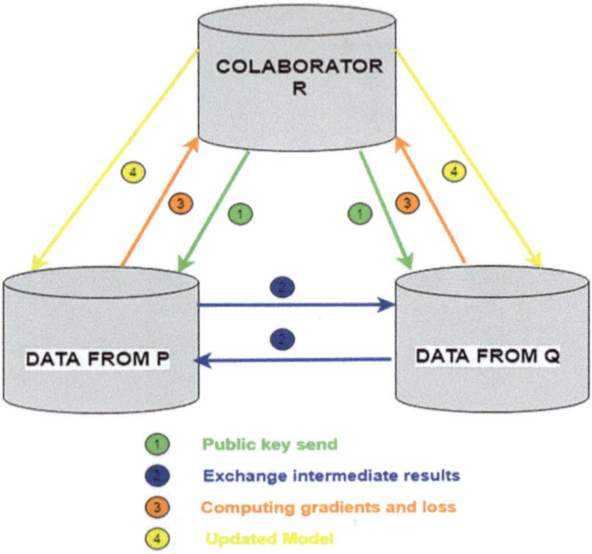

Fig. 3 Feature-based federated learning architecture

4. R decrypts and sends the decrypt gradient and loss to P and Q, and then they update the parameters accordingly.

3.3 *Federated Transfer Learning (FTL)*

FTL is a subcategory of FL that differs from both feature-based federated learning (FBFL) and sample-based federated learning (SBFL). As shown in Fig. 4, the sample space and feature space of the two datasets differ. Consider two companies P and Q due to differences in nature of the business. There is a small overlap in sample space and feature space between both companies. Adding small overlapping features and data, a model is learned from P and transferred to Q (Table 1).

To train a global model, SBFL applies to the cases of the same feature space but different in the sample, i.e., the wide overlap between the feature space. On the other hand, federated learning (FBFL) is applicable where sample space is the same but different in feature space, i.e., a wide overlap in the sample space. FTL is used when there is a small overlap in both sample space and feature space, as indicated by the dotted line in Fig. 4. To predict the labels for the target domain party knowledge are transferred from non-overlapping features from source "P" to the target domain [7].

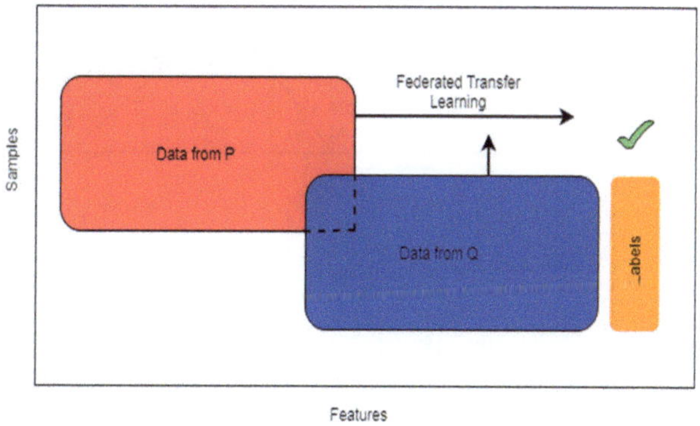

Fig. 4 Framework of federated transfer learning (FTL)

Table 1 Architecture advantages and the main focus

Architectures	Advantages	Focus on
Sample-based federated learning	Independence	Security
Feature-based federated learning	Encryption	Privacy
Federated transfer learning	Higher accuracy encryption	Prevents accuracy loss

4 Federated Learning Applied to IoT Device

The architecture of an IoT system [8, 9] consists of a hierarchical structure of three layers. The three layers are device layer, middle layer, and application layer. The IoT system with distributed learning and with FL is shown in Figs. 5 and 6, respectively. The device layer produces and collects data. It includes hardware devices that act as sensing devices. The middle layer acts as an intermediate and is responsible for transferring data from the device layer to the application layer for further processing. The application layer provides services that integrate or analyze information obtained from the middle layer.

The IoT system's middle layer links IoT devices to a cloud, such as sensors or cameras. This scenario, however, also leads to an increase in network traffic, data processing time, and the possibility of unauthorized access to data, thus adversely affecting the overall efficiency of the data analysis process. These drawbacks can be removed by FL. On the other hand, in the FL framework, the trained model is

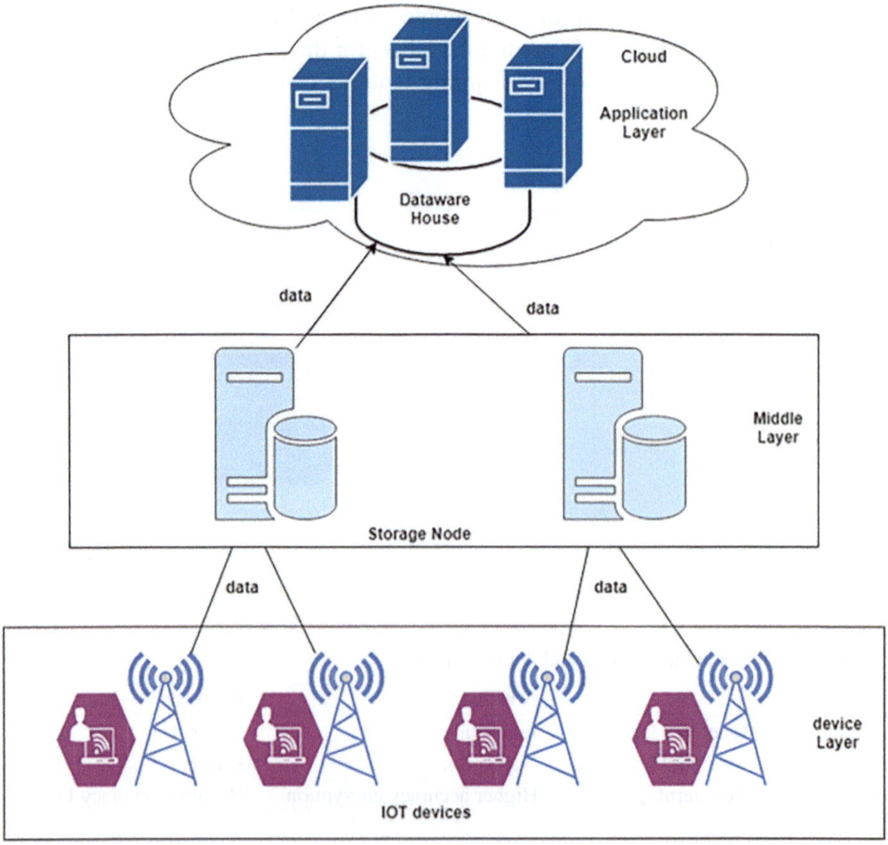

Fig. 5 Internet of things (IoT) system with distributed learning

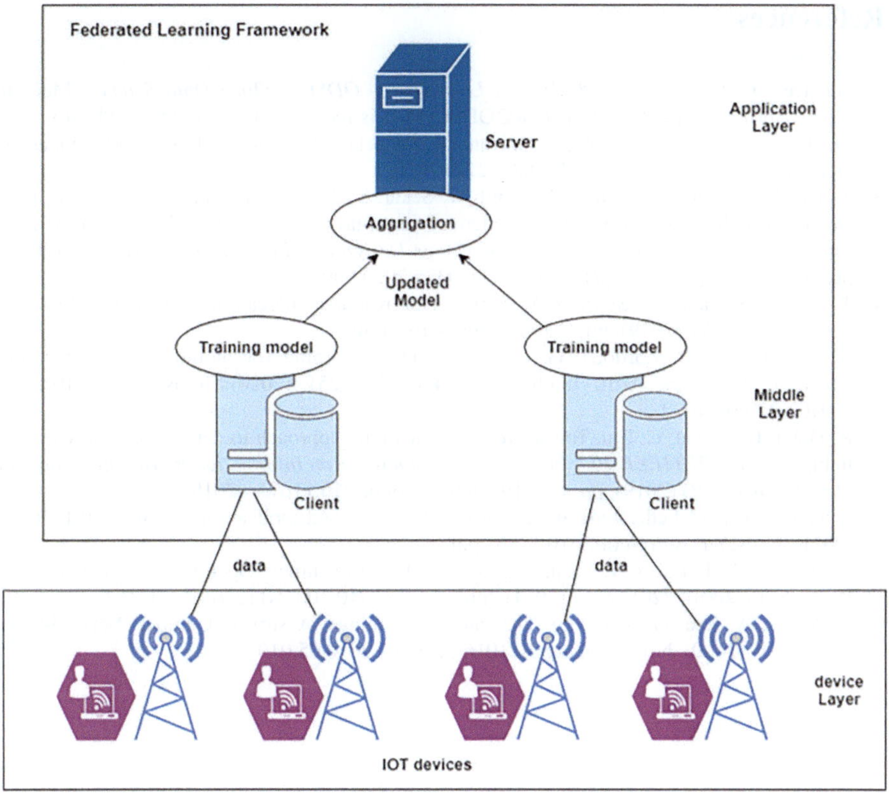

Fig. 6 Internet of things (IoT) system with FL

transmitted to the server in place of data to generate an aggregated analysis model. The updated model is then distributed to end devices, thus improving the privacy of data.

5 Conclusion

This chapter shows some observations on the behavior of FL. First of all, we start with two obstacles in AI that are distributed data and its privacy and security. FL easily overcome these drawbacks up to some extent. Two entities are there, user and server. The user trains the model on local data and sends their model to the server in place of sending the data as in traditional ML techniques. The server then aggregates the models received from different participants and then redistributes them to all users. Afterward, we provide the review on the categorization of FL based on data distribution. Finally, we discuss the Fl applied on IoT devices.

References

1. Science, O.-O. D., *What is Federated Learning? – ODSC – Open Data Science*. Medium (2020, April 3). https://medium.com/@ODSC/what-is-federated-learning-99c7fc9bc4f5
2. M. Baldonado, C.-C.K. Chang, L. Gravano, A. Paepcke, The stanford digital library metadata architecture. Int. J. Digit. Libr. **1**, 108–121 (1997)
3. T. Yu, T. Li, Y. Sun, S. Nanda, V. Smith, V. Sekar, S. Seshan, Learning context-aware policies from multiple smart homes via federated multi-task learning, in *2020 IEEE/ACM Fifth International Conference on Internet-of-Things Design and Implementation (IoTDI)* (2020), pp. 104–115. https://doi.org/10.1109/iotdi49375.2020.00017
4. T. Li, M. Sanjabi, A. Beirami, V. Smith, Fair resource allocation in federated learning. arXiv:1905.10497 (2019). http://arxiv.org/abs/195.10497
5. H. Ben Yedder, B. Cardoen, G. Hamarneh, Deep learning for biomedical image reconstruction: A survey. Artif. Intell. Rev. **54**(1), 215–251 (2020). https://doi.org/10.1007/s10462-020-09861-2
6. R. Doku, D. Rawat, C. Liu, Toward federated learning approach to determine data relevance in big data. in *2019 IEEE 20th International Conference on Information Reuse Integration for Data Science (IRI)* (2019), pp. 184–192. https://doi.org/10.1109/iri.2019.00039
7. S. Saha, T. Ahmad, Federated transfer learning: Concepts and applications. arXiv:2010.15561v2 (2021). https://arxiv.org/pdf/2010.15561.pdf
8. C.-W. Tsai, C.-F. Lai, A.V. Vasilakos, Future internet of things: Open issues and challenges. Wirel. Netw **20**(8), 2201–2217 (2014). https://doi.org/10.1007/s11276-014-0731-0
9. L. Atzori, A. Iera, G. Morabito, The internet of things: A survey. Comput. Netw. **54**(15), 2787–2805 (2010). https://doi.org/10.1016/j.comnet.2010.05.010

Federated Learning with Cooperating Devices: A Consensus Approach

Radhika Vadhi and Abhishek Sharma

1 Introduction

Cutting-edge distant affiliations are envisioned to guarantee surrender and unreasonably high strong receptiveness any spot, at whatever point, and advancing [1]. This concept will satisfy the consistent objectives for pushing toward self-controlling vehicles. To this end, On Vehicle Machine Learning (oVML) is an amazing methodology wherein each vehicle keeps up its best AI model and is relying upon the circumstance be prepared for making capable decisions, despite when it loses the network for a long time. Organizing such an oVML model could require a more conspicuous number of starter data than those per each vehicle. Fittingly, it demands data trading (and data exchanges) with interfacing vehicles [2, 4, 5]. This concept surveys the fundamental of setting up each oVML model by mistreating joined learning with their adjoining vehicles [6–10]. Without a particularly coordinated resoluteness parcel, vehicles will be hesitant to join set learning tasks, which may obstruct the social event of joined learning. One may equivalently obtain the understanding speculation approach according to Schizas et al. [11] to design a convincing force instrument for pulling in vehicles with particularly basic data to study the learning.

PC-based insight procedures in different styles have recently been applied to improve the performance of self-controlling vehicles [12]: the PC vision for slowing hindrances, AI for changing their speed to the milieu (e.g., the severity of the

R. Vadhi (✉)
Srinivasa Institute of Engineering and Technology, (Autonomous to JNTUK, Kakinada), Andhra Pradesh, India
e-mail: vradhika@sriniet.edu.in

A. Sharma
City University of Hong Kong, Kowloon Tong, Hong Kong

S. P. Yadav et al. (eds.), *Federated Learning for IoT Applications*, EAI/Springer Innovations in Communication and Computing,
https://doi.org/10.1007/978-3-030-85559-8_6

street). Considering the perhaps high number of self-driving vehicles and the necessity for these to rapidly react to authentic conditions, the conventional cloud approach creates hazards. Brought-together learning can address a response for limiting the volume of data move and stimulating such learning patterns of self-administering vehicles. The predicted result is a productive system for such an arrangement that changes our present vehicles into flexible worker ranches, performs joined learning, and reacts helpfully to their requirements. Coming benefits fuse an unrivaled plan of predictable messages and concede internet services advancing for such related self-administering vehicles.

Because the conceivably tremendous number of anticipated self-controlling vehicles and the requirement for them to quickly respond to genuine conditions, the current cloud-based learning approach is lethargic and may create perils. One huge issue is that clandestinely assembled data tests are compelled by each oVML. In this manner, their trading and data sharing should keep the unrefined data away from other adjoining vehicles. In such a way, as proposed by Google's gathered learning (GFL) [7], each oVML trades its carefully engineered model update (basically, tendency cutoff focuses and learning weightage) rather than the unrefined data. Joined learning can address a response for restricting the volume of data move and fortifying the learning measures. Also, the recovery of the unrefined data from the traded model is phenomenal, as such guaranteeing security. Such trading GFL is administered by the guide of a concentrated laborer that makes an overall model, which is an outfit normal of the generally gigantic number of clandestinely coordinated model updates. Starting there, each oVML downloads from one side of the planet to the other enabled model and registers their next domain update until the overall model is satisfied. Seeing that, considering the shut circle exchanges (a stealthily engineered model update followed by an overall adding up to show an update, setting off the next illustration of a nearby setup), the acquiescence achieved in orchestrating peak of GFL may by and large be a few minutes (in any event 10), as requested for Google's comfort application [13]. In oVML, similarly, a bound-together master may not be open.

To this end, we propose (and survey) a blockchain-based federated learning (BFL) model [3], as addressed in Fig. 1, for efficient correspondence of independent vehicles because GFL is not immediately significant considering the two basic needs.

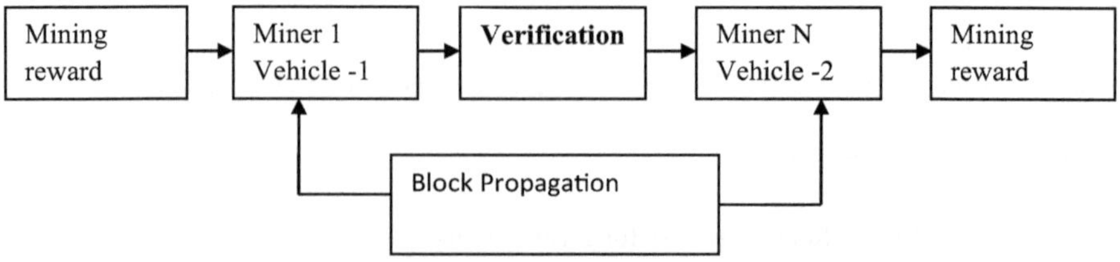

Fig. 1 Communication between components of our blockchain-based federated learning approach for sovereign vehicles

(i) Centralization: GFL depends on alone overall specialist, which is vulnerable against master disconnecting, is outstandingly in danger to organized affiliations, and encounters excited bottlenecks (resulting from the traffic of neighborhood model updates from each oVML). Such besieging for the most part makes a misguided "all things considered" model build up the response time and cause mutilation over oVML.
(ii) Loyalty: An oVML with an epic number of data tests contributes positively to masterminding for the most part. Regardless, GFL misses the mark on a framework to reward such oVML. Without an unfaltering quality program, such a vehicle with impossibly critical data tests and information will not be stimulated to consolidate reasonably with a few various tests.

To pick the two necessities of GFL actually alluded to, we influence blockchain [14–20] (swear off the joined as a rule master of GFL and use a blockchain) and propose a blockchain-based federated learning (BFL) approach, where the alliance system interfaces with exchanging nearby model updates from vehicles while giving and checking their related rewards [21]. Figure 1 portrays our proposed BFL, which applies to controlling vehicular connection structure by vanquishing the assembled assaulting issue and broadening the level of its interest to contriving vehicles in a public relationship by an endorsing illustration of the space-organizing modules. Significantly more, all things considered, by giving prizes such as the accommodation of data test sizes, BFL upholds vehicles with the greater size of data tests to join set learning. A hypothetical point of view on the proposed BFL framework including self-controlling vehicles and diggers is shown in Fig. 1. The farm trucks can be either genuinely moving vehicles or separate concentrations at network edges (for instance, cell base stations), which are generally computationally basic for the mining interaction. We propose a uniform strange vehicle ranch truck affiliation plot. Vehicle ranch hauler affiliation using the uniform discretionary strategy (as its wellspring of an affirmation) in their organization is thus trusting in either the diggers or the oVMLs of our BFL structure; the two of which are routinely dim sections to each other. The standard benefits of a discretionary vehicle–tractor association course of action is as per going with the following:

(i) An enemy who can oversee diggers in other fixed affiliation plans cannot abuse this connection construction to affect which oVMLs it wants to settle.
(ii) It defines a game on the blockchain that helps its players with playing carelessly, depends on the wake of pushing powers, and does not take any considered their authenticity.
(iii) It is more straightforward to shape delegate packs from the overall vehicles in the BFL structure.
(iv) It allows the overall model and findings to apply to the entire BFL system.
(v) It is clear, computationally unassuming, and diminishes the risk of knowing or thoughtless propensity. The organized endeavors in Fig. 1 are explained as follows. Each oVML performs nearby learning and sends the local model update to its related earthmover in the association. All farm haulers exchange and admit the whole of their recent model updates, and now and then later execute

their affirmation of work [22]. when a digger finishes its attestation of work, it makes a square exactly by recording the upheld nearby model updates. In the end, the square (passed on the block including undeniably the nearby model updates) is inserted into the scattered record of the blockchain. This actual implanted square would then have the choice to be used by totally related vehicles to handle the importance in an overall model update.

2 Blockchain

The absence of versatility is known to be the preeminent deterrent stopping the traffic of mass reception of blockchain innovation. All current blockchain projects search for arrangements that could improve the exhibition of their organization. Many arising projects guarantee that they have a wizardry shot that could tackle the issue. In any case, such affirmations are not generally substantial. Sadly, numerous onlookers and speculators do not understand the center and base of this issue. Without profound examination and significant specialized foundation, it is hard to decide shrouded bottlenecks and compromises. In this, we will talk about one notable bottleneck that keeps Bitcoin from scaling. Soon after the innovation of decentralized distributed organization Bitcoin, specialists became eager about what decides the restrictions of Bitcoin's scaling. Before long the center issue was resolved and portrayed regarding block proliferation time or square-engendering delay. It is a normal time that is required for the new square to arrive at most of the hubs in the organization. In an enormous decentralized organization such as Bitcoin, at whatever point the new square is produced, it is communicated by the gossip convention. In the event that some hub has the new legitimate square, it educates hubs associated with it about its new belonging.

At that point the hub moves this square to those hubs which have requested that it do so. Before the square arrives at each full hub in the organization, it goes through seven go-between hubs. It is significant that each legitimate hub checks the square before transferring it to different companions. Clearly, the entire process takes some time. Each new square shakes the organization and makes hubs and ethernet associations between them work at full force.

One may contend that since the dispatch of the organization, there have been numerous upgrades to the gossip convention. For instance, the Bitcoin improvement proposition BIP 0152 acquainted the alternative with moving just short exchange IDs, rather than the entire rundown of exchanges, in the square body. Notwithstanding, if the hub does not have that exchange in its mempool, it needs to request that its companions move it in a different message. On the off chance that there is an enormous number of such exchanges in the square, at that point improvement from BIP 0152 vanishes. Because information transmission is the most tedious piece of the square handoff, analysts became eager to decide how long is needed for an information parcel of a specific size to arrive at 50% (half), 90%, or 95% of hubs in the organization.

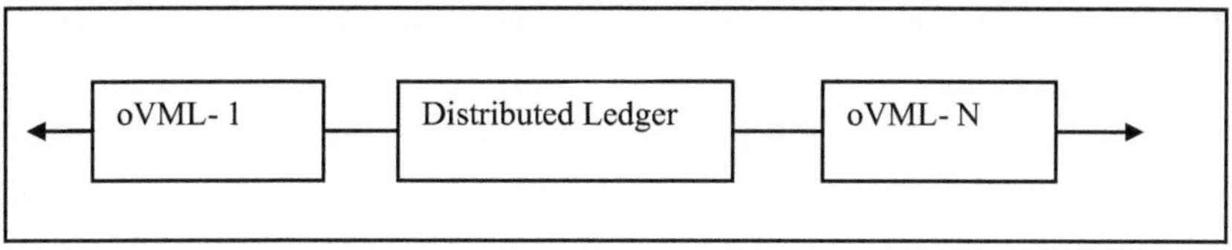

Fig. 2 Verification

It was discovered that for squares larger than 20 KB, the square-engendering delay almost corresponds to the square size. As indicated by research published in 2013, each additional kilobyte of information in the square caused an extra 80 ms of square-engendering delay. From that point forward, several scholarly papers and reviews on this subject have been distributed each year. They update the previously mentioned information and talk about different improvement propositions. Also, the site screens the present status of the Bitcoin organization and the proliferation time. Additionally, it furnishes diagrams in Fig. 1 with a chronicled information square regarding this matter. The verification is explained in Fig. 2. Most spread time massively affects blockchain security. The more drawn out the engendering time in the organization, the more regularly excavators mine on top of old squares. Therefore, the forking of the primary chain happens all the more frequently, and the level of what could be a vagrant square rise. The long extended postponement prompts the purported verifiers dilemma. A few hubs may find that grounded blockchain networks have a similar plan as Bitcoin. Accordingly, the square proliferation time in these organizations complies with similar principles. Lamentably, the square skirting the square check step is a beneficial procedure. For this situation, they face the danger of mining on top of some unacceptable square. In any case, if block check time is critical, this methodology could be beneficial. Analysts found that long proliferation delay lessens the hub opposition against 51% assaults and childish mining. To address the previously mentioned issue, blockchain designers regularly attempt to keep the square engendering time to be less than 1% of the normal square time: this is valid for Bitcoin, Ethereum, and other major blockchain networks that depend on PoW agreement. Therefore, block engendering time to half the hubs in the Bitcoin network is regularly less than 6 s. It is significant that even in the direst outcome imaginable, the engendering postponement ought to be sensible so excavators will keep their hubs synchronized more often than not, and will consistently check proposed blocks.

At whatever point individuals talk about the versatility of the blockchain, they notice the exchange throughput of the framework. Notwithstanding, individuals failed to remember that enhancements in exchange throughput should not bargain the organization's security, or raise information stockpiling prerequisites for hubs craving to take an interest in the organization. These adjustments could diminish the number of free-exchange validators in the organization, in this way lessening decentralization.

The Bitcoin throughput transaction can be determine by using this formula:

$$\text{Throughput} = \frac{B_S}{T_S * B_t}$$

where B_S is the block size in bytes,
T_S is the average size of transaction in the block
B is the average time between consecutive blocks in the blockchain.

Clearly, exchange throughput could be expanded by expanding the square size, by decreasing the exchange record size, or by diminishing the span between blocks. It is somewhat difficult to lessen the size of the exchange records. One may rather attempt the other two choices. In any case, these activities will build the level of time that is spent on square proliferation. Accordingly, the security and decentralization of the organization could be bargained.

It may be that network assets are utilized wastefully in the portrayed Bitcoin convention. Each hub measures and sends the essential information about another square just a small part of the time. Its organization data transmission is truly significant; however, it is utilized in full for a couple of moments all at once. The remainder of the time, this hub communicates just forthcoming exchanges and helper information. This perception has motivated analysts to search for more productive convention plans that could significantly improve exchange throughput without trading off the security and decentralization of the organization.

With BFL, note that the worldwide model update is locally calculable at every vehicle. Also, the BFL configuration additionally ensures that the breaking down or disappointment of a digger, hub, or vehicle antagonistically affects the worldwide model updates for some other vehicles (in general, the organizational framework). Be that as it may, there is a tradeoff: to understand the entirety of the previously mentioned benefits, instead of the GFL, BFL has an extra deferral because of the blockchain of the executives in the framework. To evaluate (and address) the persisted delay by blockchain, the total start-to-finish postponed examination is directed for the framework with BFL by representing the confirmation of work, correspondence, and calculation delays. With applicable experiences from the model, we additionally limit the by and large postponement of the framework by powerfully changing the square appearance rate, that is, the intricacy of verification of work for the blockchain framework.

In light of the foregoing discussion, the principal commitments are as follows.

1. To propose and build up a novel and completely decentralized blockchain-based unified learning structure to guarantee start-to-finish reliability and postponement for the impending self-governing vehicular systems administration framework.
2. To build up a complete numerical investigation of the framework elements for the start-to-finish defer examination, which gives significant knowledge about limiting deferral and determining ideal square appearance rate by considering correspondence inertness and agreement delays throughout.

3. By organizing a comparable to vehicular organization model, which consolidates the mind-boggling associations of the remote channel and connection attributes with the definite transmission component of the phone organizations, it prevails with regard to estimating the effect on the presentation of our BFL framework. We perform broad mathematical and reproduction analyses to assess the proposed structure under different organizational situations and channel conditions.

   Because of the time-fluctuating nature of the self-governing vehicular organization settings, for example, the quantity of vehicles, excavators, and remote direct conditions differ progressively in the organization situation. Thus, to push a step ahead, plan an online calculation for limiting the framework delay under reasonable organization conditions.
4. In view of the closed structure articulation determined utilizing the created scientific model under static organization conditions, a constant online calculation that ceaselessly screens the deviation in the framework delay from a focus on esteem and adjusts the square appearance rate to guarantee insignificant total postponement and adapt perfectly to the transient organization elements.

All the more critically, with pertinent understanding from this work, various difficulties (gaps in information) should be settled to understand these changes. Specifically, we distinguish the accompanying three-key bottleneck difficulties requiring further examination and give possible future headings.

- Advancement of refined convenience models considering specific features of self-administering vehicles for passed-on learning.
- Plan of threat examination and control draw near for security spillage during verification.
- Analysis of setting carefully joined (and generous) circled learning-based blockchain system for very strong low deference applications.

3 Agreement-Based United Averaging

The proposed approach disassembled here licenses the devices to obtain capacity as far as possible, by depending only on close (in-network) interaction with and composed exertion with neighbors. The geography of the satisfying affiliation has appeared as a coordinated blueprint $G = (V, \xi)$ with the course of action of focus $V = \{1, 2, ..., K\}$ and edges (joins) ξ. As portrayed in Fig. 3, the K contraptions are connected through a decentralized correspondence plan dependent on gadget-to-gadget (D2D), or structureless, correspondence. The neighbor set of gadget k is exhibited as $Nk = \{i \in V: (I, k) \in \xi\}$, with cardinality $|Nk|$. The set $Nk^{-} = Nk \setminus \{k\}$ contains only the neighbors to k. As beforehand, every gadget has a data set of models that are utilized to set up a neighborhood NN model $W_{t,k}$ sooner or later t. The input model assistants highlights x into yields $\hat{y}(W_{t,k}; x)$, and a non-blended expense work Lk $(W_{t,k})$ is utilized for the improvement of $W_{t,k}$.

Fig. 3 Federated learning

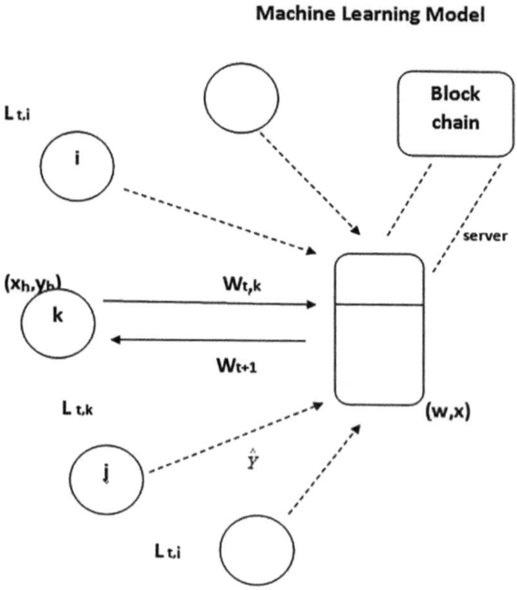

The use of combined streamlining to decentralized organizations with subjective network formats makes the issue more difficult with contrasted and customary brought-together gadget alliance, depending on an immediate connection with the worker. Agreement-based unified averaging (CFA) is intended to completely or halfway supplant the concentrated combination performed by the worker in federated average (FA), for example, at the point when the correspondence with the worker is not accessible [17]. Introduction of $W_{t,\ k}$ is first acquired at time $t = 0$ (when the agreement begins) and may be additionally performed in this manner from a model update obtained from the worker when running FA. As portrayed in Fig. 4, at each new correspondence round ($t > 0$) the devices perform model conglomeration utilizing the model updates $W_{t,\ i}$ procured from the neighbor gadgets $I \in N_k$.

$$\psi_{t,k} = W_{t,k} + \in_t \sum_{i \in N_K} \sigma_{k,i} \left(W_{t,i} - W_{t,k} \right) \tag{1}$$

where $\in_t$ is the consensus step-size and $\sigma_{k,\ i}$ are the mixing weights for the models that are chosen as

$$\sigma_{k,i} \frac{E_i}{\sum_{i \in N_k} E_i} \tag{2}$$

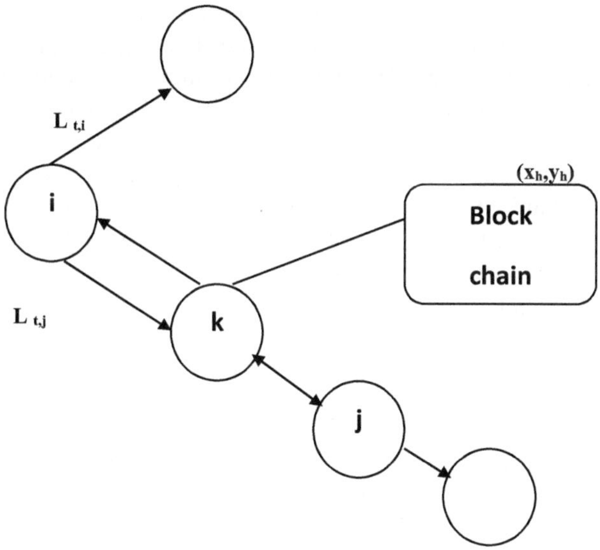

Fig. 4 Distributed/consensus-based federated learning (FL) (server based)

Model adaptation is performed:

$$W_{t+1,k} = \psi_{t,k} - \mu_t \nabla L_{t,k}\left(\psi_{t,k}\right) \tag{3}$$

by running stochastic gradient descent (SGD) enhancement w.r.t. the collected model $\psi_{t,k}$ and over various smaller than expected clusters of nearby prepared information. The size B of smaller than normal clusters is with the end goal that B < E_k, whereas various decisions can be chosen. As in average SGD, the progression size μ_k can be tuned on continuous ages t to improve union [16]. Model collection (Eq. 1) resembles the entirety of the weight tattle convention [18, 19] when setting $t = 1$. The progression size t controls the agreement security. Following SGD transformation (Eq. 3), the model update $W_{t+1,k}$ is then sent to neighbors $N_{\bar{k}}$ for another agreement round. Different decisions for blending loads in Eq. (2) can be founded on weighted agreement [12] or dispersion techniques [20]. The agreement step-size t can be chosen as in Jeong et al. and Olfati-Saber et al. [10, 12]. The exchange among μ_t and agreement t for combination is handled as by Cattivelli et al. [17].

4 Solidifying Federated Averaging with Consensus. A Case Study: A Contextual Analysis

The reference circumstance for use of the proposed in association FL contraptions incorporates a thick relationship of free Internet of Things (IoT) difficult-to-reach devices [21] that are perceiving their normal fragments using frequency-modulated continuous wave (FMCW) radars [22] working in the 122-GHz band (see Fig. 3, top). Radars in the mm-wave (or sub-THz) groups are generally feasible in mechanical creation lines (or computerized cells) for speed–distance examination and considerably more generally expanded reality applications. The contraptions screen a conventional current workspace during human–robot collaboration (HRC) attempts [23] to isolate body upgrades from robots and, thus, to see the distance of the expert from the robot and potential perilous conditions [24]. The ML model is here set up to pack C = 8 potential HRC conditions depicted by different HR distances (up to 4 m) standing apart from hazardous conditions [17]. As a result of their colossal size, the data accumulated by the radars cannot be reliably dispatched back to the master for organizing or recalibration, as the correspondence related to the specialist is perhaps similarly not open or overwhelmingly stacked [25] to ensure robot reevaluation for worker prospering. Cross-territory transparency is accordingly used here as a replacement to consolidated connection structure whereas the proposed in-network FL mechanical social affair fills in as a replacement to master side figurations. The dataset (Ascorti et al. [26]) fuses beat signal spectra, changed over using 512-point fast Fourier transform (FFT): move speed is 6 GHz, carrier reiterate 119 GHz, slant (beat) length of $T = 1$ ms. The model information Ek is many Ek = 10 FFT tests xh of evaluations 512 × 1, h = 1, ..., Ek, amassed unreservedly by single devices and put to the side locally. Notice that most of the contraptions have neighborhood models only for a subset of the C = 8 classes of the overall model. Minimal get-togethers for neighborhood penchants have a size comparable to B = 5; orchestrating passes in this course incorporate Ek/B = 2 more unpretentious than anticipated gatherings whereas datasets consolidate E = 16,000 assessments. The ML in the overall model is depicted by a NN with Q = 2 useful layers: convolutional and totally related, as noted in the upper-right corner of Fig. 4. The looking at loads and propensities W = w0T,1, w1,1, w0T,2, w1,2 have evaluations identical to d1 = 16, d2 = 8 for utilitarian layer 1 and d1 = 168, d2 = C for layer 2. Nearby model updates (for each layer) are delivered off neighbors by using a time division multiple access (TDMA) medium. The dataset (Ascorti et al. [26]) includes beat signal spectra, changed over utilizing 512-point fast Fourier transform (FFT): move speed is 6 GHz, with transporter rehash 119 GHz, incline (beat) length of $T = 1$ ms. The model information is a ton of = 10 FFT tests xh of assessments 512 × 1, $h = 1, ...,$ aggregated independently by singular contraptions and put to the side locally. Notice that the majority of the contraptions have nearby models just for a subset of the C = 8 classes of the general model. More unassuming than expected packs for neighborhood inclines have a size indistinguishable from B = 5; preparing passes hence incorporate/B = 2 small gatherings while datasets include = 16,000 appraisals.

The ML in the general model is portrayed by a NN with Q = 2 serviceable layers: convolutional and completely related. This appears in the upper right corner of Fig. 4. The relating weights and tendencies W = w0T,1, w1,1, w0T,2, w1,2 have assessments equivalent to d1 = 16, d2 = 8 for useful layer 1 and d1 = 168, d2 = C for layer 2. Nearby model updates (for each layer) are shipped off neighbors by utilizing a time division multiple access (TDMA) medium sharing correspondence plot. Execution is explored with respect to underwriting calamity, considering the full support dataset. Cross-entropy difficulty is utilized for point assessment sharing correspondence plots. Execution is investigated concerning underwriting difficulty, considering the full support dataset.

Cross-entropy difficulty is utilized for point assessment log [(Wt,k; xh)] while getting together and execution have been rehashed on a virtual climate that makes a discretionary number of virtual contraptions (and worker). The thing, written in Python, utilizes tensor flow backend and multiprocessing libraries: simplified configurations for testing CFA approaches are given in the vault [26], near to the dataset utilized.

For diversions, we think about a k-standard relationship of $K = 80$ contraptions: all gadgets have a tantamount number of neighbors or degree. In Figs. 3 and 4, we consider the support difficulty separating worker-less CFA (where expert correspondence is utilized especially for generally model show W0,k), bound together FA (where region models are joined at the trained professional), and cream methods that other Ns rounds of joined FA with Nc rounds of course of action based CFA, expecting that correspondence with the worker is open irregularly, as exhibited by the breaking point α = Ns/Nc. Specifically, during CFA announcement, at time t = t0, every contraption utilizes the last accessible general model obtained from the subject matter expert, Wt0,k = WNs, for example after the Ns correspondence rounds of the past FA stage, and gets a nearby update through SGD:

$$W_{t+1,k} = \mathrm{WNs} - \mu_{\mathrm{t}} \nabla \mathrm{L}_{t,k}\left(\mathrm{WNs}\right) \tag{5}$$

where μs is the SGD step size and $\nabla \mathrm{L}(W_t) = \nabla W_t\,[\mathrm{L}(W_t)]$.

The incline of the mishap in Eq. (3) over the named clusters and w.r.t. the model Wt. Backpropagation is utilized here for inclinations assessment. The model measure at the social gathering is appeared as W = limit→∞ Wt. Rather than giving the available information to the subject matter expert, in FL as far as possible W are updated aggregately by interconnected contraptions, going likely as nearby understudies. On each round t of correspondence, the expert disperses the current by and large model Wt to a subset St of nt gadgets. The gadgets fully update the model Wt utilizing the focuses (SGD) from nearby arranging information as

$$W_{t+1} = W_t - \mu \times \nabla L_{t,k}\left(W_t\right) \tag{6}$$

Where $\nabla \mathrm{L}_{t,k}$ (Wt) = ∇W [$\mathrm{L}_{t,k}$ (Wt)] addresses the inclination of the misfortune (Eq. 4) seen by the k-th gadget and w.r.t. the model W_t. Updates $\nabla \mathrm{L}_{t,k}$ (6), or neighborhood models $\mathrm{W}_{t+1,k}$, are sent back to the worker, after quantization, anonymization [7], and pressure stages, displayed here by the administrator PΘ. A worldwide model update is acquired by the worker through conglomeration as per

$$W_{t+1} = W_t - \mu_s \frac{1}{nt} \sum_{k=1}^{n_k} \frac{E_k}{E} \rho_{\Theta}[\nabla L_{t,k}\left(W_t\right) \tag{7}$$

This resuoy is managed back to neighbors to begin understanding cycles containing the blend in (Eq. 5) and assortment (Eq. 7) stages. For CFA, we pick $t = 0.5$, $\mu t = 0.025$, and, as far as possible, σk,i as in (Eq. 6). The two wonderful attributes $\alpha = 0$ and $\alpha \to \infty$ consider, autonomously, to worker-less CFA (Ns = 0) and concentrated FA (Nc = 0).

In Fig. 3, we contrast the rotated FA-CFA system and $\alpha = 1$ and Ns = Nc = 10, with the two reference settings (i.e., FA and CFA). Underwriting occurrence decreases over steady occasions of the subbed FA-CFA technique to the degree the model updates Wt,k combine to the genuine by and large model W∞. Considering CFA, model updates are as of now settled distinctly on participation with neighbors; consequently, affiliation is much slower separated and FA (ran lines). Precisely true to form, affiliation improves with the number of neighbors, from |Nk⁻ | = 2 to 4. Notice that running SGD on neighborhood information is only (withdrawn preparation); thus, debilitating plan and CFA would result in overfitting and a somewhat more prominent underwriting catastrophe.

With the occurrence to information and correspondence advancement, it has become indeed less difficult to accumulate a very large amount of data, and thusly, data-driven procedures are attracting exceptional interest from both industry and the academic world. Standard enormous data-based AI systems accumulate data to a particular territory, similar to the central laborers. In any case, of late, security has been potentially the principal stresses in the wide association of these gigantic data stages, which is seen in Fig. 3.Typically, data have a spot with different social events, which may disregard exchanging data among themselves on account of safety impediments, similar to the general data protection regulation (GDPR) [1].

In the Internet of Things (IoT) period, client gadgets produce much information that can be utilized to improve the client experience of construction. In any case, clients are hesitant to give their own information, considering the hazard of information abuse and spillage. Then again, some IoT applications in vehicular conditions, including charming driving, should settle on ideal choices subject to various types of vehicle sensor information, including the Global Positioning System (GPS), camera, and radar. Also, there are two rules to contemplate. From the outset, although the cloud can manage information and settle on choices subject to general data, amassing and moving information from specialists to the cloud requires high trade speed and causes a high deferral, moreover, while the streamed topic specialists. To settle the issues referred to of late, united learning (FL) has been proposed by Google [2] to permit different parties to regularly set up a model, which includes neural affiliation limits while relieving the confirmation risks. In Savazzi et al. [2], different customers (or laborers) help the focal worker to set up a huge neural affiliation model. The focal worker first spreads an essential model of the course of action to the customers. Considering the model, every customer figures its nearby updates of the general model, as stochastic inclination drop (SGD), considering its own neighborhood dataset. After a predefined arranging period, all customers send their own updates (SGD) to the focal trained professional, and the focal worker collates these updates to calculate a general model. Joined averaging (FedAvg) assessment knows about hard and fast nearby invigorates made by the customers. These techniques are rehashed until the focal expert accomplishes a satisfactory by-and-large model. Because all neighborhood datasets amassed by the customers are not moved and handled at the focal worker, information protection is developed.

Not identical to standard decentralized learning that moves close, FL is expected to accomplish key central focuses:

- Better fitting for non-IID flowed and conflicting information: FL unwinds the idea of self-administering and enigmatically passed on (IID) information. While existing decentralized methods of reasoning essentially expect an IID portion of information among preparing to topic specialists, FL can accomplish better execution in non-IID information by efficient customer affirmation and utilizing grouping calculations. FL is additionally furnished for overseeing conflicting information dispersing. Various customers could have various sizes of information with various degrees of significance. FL, which depends upon a free relationship of customers and the authority of the trained professional, prompts an efficient treatment of lopsided information.
- Low correspondence overhead: FL diminishes the size of correspondence information just by sharing the neighborhood updates of IIOT.

Bound together changing consistently applies when specific entertainers need to design models on more prominent datasets than their own, yet cannot tolerate giving the information in itself to others (e.g., for veritable, key, or cash-related reasons). The turn of events yet requires an unbelievable connection between nearby workers and the most non-computational force for each node [27].

4.1 Transportation: Self-Driving Vehicles

Self-driving vehicles represent different AI kinds of progress to work: PC vision for researching deterrents, and AI for changing their speed to the environment (e.g., the awkwardness of the road). Considering the conceivable high number of self-driving vehicles and the fundamental for them to quickly respond to genuine conditions, the regular cloud approach may take risks. Bound together learning can address and respond in due demand concerning restricting the volume of data move and vivifying learning measures [28, 29].

4.2 Industry 4.0: Shrewd Hoarding

In Industry 4.0 there is an incredible decision of AI methods to improve the productivity and appropriateness of the mechanical cycle while ensuring a specific degree of security. Considering everything, the security of delicate information for experiences and gathering affiliations is of central significance. Joined learning calculations can be applied to these issues as they do not uncover any fragile information [30].

4.2.1 Medication: Advanced Flourishing

Merged learning intends to address the issue of information association and confirmation by methods for arranging assessments appropriately without trading authentic information. The current standard methodology of joining information from various focuses goes to the impairment of fundamental concerns concerning driving forward security and information assurance. To handle this issue, the ability to plan AI models at scale across various clinical establishments without moving the data is a basic function. *Nature Digital Medicine* published in September 2020 a paper entitled "The Future of Digital Health with Federated Learning" [31] in which the maker's assessment of how joined learning may offer a response for the possible destiny of mechanized thriving and highlight the burdens and considerations that ought to be considered.

We consider a streamed cell cloud (MDC) climate, which contains two substances: (i) specialist gadgets (i.e., laborers) and (ii) purchaser contraptions (i.e., purchasers). The purchasers having an association-concentrated undertaking select computational asset redesign laborer gadgets to achieve these assignments. The correspondence between the specialist and purchaser contraptions is engaged through Wi-Fi sections or Bluetooth.

Let $b = \{b1,2, \ldots, bm\}$ infers the arrangement of m purchaser contraptions, each having one register concentrated errand. In this paper, we used trials and purchasers correspondingly. Executing these assignments requires higher attention to force also to an alternate extent of assets, that is, CPU cycles and memory because of the heterogeneous thought about those undertakings. The computational undertaking of every purchaser, $bi\in$ b, is credited by its asset fundamental, task size λ, task defer cutoff time, and genuine valuation vi, keeping track of the extent of cash that can be paid to a prepared proficient, and this worth relies on the computational eccentrics of the assignment. We recognize that there are n asset advance master gadgets in the framework whose set is shown by w = $\{w1,2, \ldots, wn\}$. Every specialist gadget $w\in$ has Rj units of accessible assets with the least. Regardless, our design can additionally work with different sorts of assets or blends of those. Considering versatility, master gadgets can move to another zone proficiently. From this time forward, laborers' accessibility acknowledges a tremendous part in task finish and affects the purchaser's quality of service (QoS). A laborer gadget, even with adequate assets, may give the computational outcomes exactly as expected in the event that it is over issue with undertakings, which thus would destroy the standing level of a specialist to the purchaser device. Accordingly, we limit the best extra positions waiting be finished by $njmax$. In FEST, a specialist gadget is paid by the possibility of its finished undertakings. Upon profitable fulfillment of the relegated assignments, a laborer receives its guaranteed cost. Furthermore, if the master finishes the undertaking with uncommon quality, a motivation or prize is given to the client. Here, motivation exhibits the money-related worth-added with the specialist's avowed expense to yield its part. Along these lines, the bit is the summation of the specialists' guaranteed cost and the motivation total. Obviously, a specialist is repelled for giving low undertaking quality. Considering everything, the master parcel is not just comparably much as its guaranteed cost. In the FEST structure, both the purchasers and laborers are nonvaluable as they try to extend their utilities. Notwithstanding, the purchasers basically participate to handle the leftover specialists. Figure 2 portrays the interests between the purchaser and laborer contraptions in a passed-on MDC climate. The arranged advances are recorded under these stages.

Stage 1: Each purchaser contraption $bi \in$ b from the outset submits offers to the topic specialists $wj \in$ w, each implied by ci, nearby its asset need and undertaking surrender cutoff time.
Stage 2: After getting the offers, every master contraption (i.e., sales rep) runs FEST-Auction to pick the up-and-comer purchaser gadgets and their expressed cost for executing those undertakings.
Stage 3: The notice is moved off the purchaser's gadgets about the successful offer competitors near the pronounced cost.
Stage 4: The purchaser contraptions by then distributively select the most reasonable master device dependent on its asserted regard, task finishing time, and notoriety-based utility.
Stage 5: After that, every purchaser sends the choice notification to the master nearby the computational errand.
Stage 6: After finishing the errand execution, the specialist sends back the computational outcome to the particular purchaser gadget.
Stage 7: The purchaser then overviews the computational outcomes sent by the specialist gadget, picks divide envisioning a sparkle framework, to be express FEST-Payment, and updates its standing. The significant documentation to depict the system is recorded in Table 1.

Zero-Knowledge Proofs (ZKPs) Zero-knowledge (ZK) validations are a cryptographic hash that engages one social gathering (called the prover) to show explanations to another gathering (called the verifier), which depend on limited information known to the prover, called the passerby, without revealing those upheld experiences to the verifier. Nowadays, ZKP shows can achieve proof sizes of hundreds of bytes and verifications of the mentioned for milliseconds paying little attention to the size of the affirmation being addressed.

A ZKP has three striking properties: top (if the check is real and the prover and verifier follow the show, the verifier will see the proof), ampleness (if the affirmation is fake and the verifier follows the show, the verifier will pardon the statement), and zero data (if the declaration is huge and the prover follows the show, the verifier will fundamentally find that the declaration is genuine and will not take in any confidential information from the association).

Past these normal properties, there are different sorts of zero-data upgrades for kept up language for the affirmations, outline necessities, prover, and verifier computational efficiency, between progress, conservativeness, and head hardness speculations. There are diverse ZK improvements that help specific classes of explanations, Schnorr and Sigma shows are cases of such broadly used shows. Although such shows have different uses in specific settings, general ZK structures that can keep up

Table 1 Notations

Notation	Description
B	The Set of all buyers in the system
W	The Set of all workers in the system
B'_j	The set of all winning buyers of worker $w_j \in W$
r_i	The amount of resource required of buyer $b_i \in B$
R_j	The amount of resource available at worker $w_j \in W$
r_i^d	Task deadline of buyer $b_i \in B$
r_i^c	Task completion time of buyer b_i from final worker
T_j	Availability time of worker $w_j \in W$
p_i^b	Winning price of buyer $b_i \in B$
$p_{i,j}^s$	Winning payment of worker $w_j \in W$ from buyer $b_i \in B$
$P_{i,j}$	Final price given by buyer $b_i \in B$ to worker $w_j \in W$
$c_{i,j}$	Unit resource bid submitted by buyer b_i to worker w_j
A_j	Minimum ask of worker $w_j \in W$
$\Psi_{i,j}$	Priority factor of task b_i to worker w_j
$U_{i,j}^C, U_{i,j}^T, U_{i,j}^O$	Buyer's utility due to cost, time and quality, respectively
ε_{ij}	Energy required to execute task b_i by worker w_j
$\alpha_{i,j}, \beta_{i,j}$	Number of positive and negative interaction between buyer b_i and worker w_j
ρ, η, κ, ζ	Weight of positive, negative, recent and past interactions, respectively
$\Omega_{i,j}^{fin}$	Final reputation value of buyer b_i for worker w_j

any steadiness give an essentially more widely fitting contraption (exploring for the setting of FL), and appropriately we circle around such improvements for the rest of the discussion.

A fundamental explicit portion between different movements is fundamental for trust in a system. Some ZKPs rely on a normal reference string (CRS), which is figured using special experiences that should remain covered to guarantee the ample properties of the proof. The check of such a CRS is proposed as a trusted in technique. While this need is a soft spot for such developments, the current ZKP improvements that achieve most short affirmations and verifier's efficiency require trust in a game plan.

Another significant property that impacts the correctness in different conditions is whether making the assertion requires an association between the prover and the verifier, and here we see nonintuitive zero-knowledge affirmations (NIZKs) that attract the prover to send a lone message to the verifier and require no further correspondence. Dependably we can change over normally to noncrafty affirmations making more grounded contemplations about the ideal utilization of hash limits (i.e., those hash limits carry on as eccentric prophets).

What is more, there are different assessments for efficiency of a ZKP structure that one should consider, for instance, the length of the confirmation and the computation multifaceted design of the prover and verifier. The ideal prover's flightiness should be straight in the execution time for the surveyed steadiness, yet many existing ZKPs present extra (from time to time significant) overhead for the prover. The most efficient verification complex plan needs to incorporate at any rate straightness in the size of the duties regarding the outlined usefulness, and in the setting of attestations made by the FL master for this information, the size will be significant.

Restricted non-sharp zero-data authentications (SNARKs) are such a ZKP that gives unsurprising proof size and verification that depends basically on the information size, straightly. These attractive efficiency properties do go to the impairment of more grounded speculations, which is by and large unavoidable, and trusted in methodology everything thought about existing approach. Most existing SNARK movements influence quadratic numerical endeavors and are right as of now open in open-source libraries; in particular, the prover incorporate ought to be straight in the size of the circuit for the demand being illustrated. Recently, a ZKP structure that achieves straight prover complex nature regardless with a consensus social occasion of upgrades that achieve a wide level of efficiency bargains, avoid the trusted in-approach basic, and use more standard cryptographic doubts. In the more recent years, extending proportions of critical applications have been using nonnormal zero-data affirmations, overall, blended by blockchains. Using basic ZKP plans and NIZKs efficiently with respect to FL remains a hazardous open request. In such a setting, NIZKs may engage to show the master properties about the client's data sources. In the setting where the verifier is the client, it will try to offer a strong articulation to check as it unites obligations from various clients. Of interest in this setting, late work attracts to manage the circumstance where the distinctive verifiers have sections of the assertion, s-based FL: a from-the-beginning model.

We inspected trusted in execution environment and inaccessible endorsement (TEEs) in this section, yet move here around the way that TEEs may offer opportunities to give verifiable checks. Doubtlessly, TEEs attract to give statements concerning and check the code (twofold) running in its present condition. In particular, when the verifier knows (or can reflect) which twofold should run in the acquired territories, TEEs really need to give a pondered legitimacy (the code execution cannot be affected by some different methods than the wellsprings of data), and certification (the TEE can show that a specific twofold is executing). Fundamentally more, considering everything, distant certification allows a verifier to securely measure inside a state of eliminated hardware organization and can be used to develop a static or dynamic establishment of trust. While TEEs interface with hardware-based distant validations, both programming-based eliminated confirmations and hybrid distant check plans were proposed in the course of action and enabled to manage gear essentials for verifiability.

In an amassed picking disturbance, TEEs and inaccessible supports may be particularly valuable for clients to have the choice to efficiently attest key cutoff focuses running on the specialist. For example, a protected blend or shuffling could run in TEEs and would give differential security guarantees on their yields. Consequently, the post-getting ready reasoning in like manner applied by the master on the differentially private data could run on the specialist and stay illegible with respect to the clients. Note that such a development strategy requires the clients to know and trust in the particular code (same) for the essential abilities to be applied in the areas. Additionally, distant assertions may enable a specialist to demand specific rudiments from the clients pulled in with the FL evaluation, similar to insufficiency of openings, ceaseless quality, and unobstructed capacity.

4.3 Affirmations Against External Malicious Actors

It is usual that the presence of a confided-in trained professional and conversation about different difficulties and open issues toward accomplishing cautious affirmation ensures against outside compromising entertainers (for example, not all-around orchestrated customers, ill-disposed subject matter experts, or severely masterminded contraptions that gobble up the learned model, or any mix thereof).

As the malevolent customers can review all messages received from the subject matter expert (checking the model repeats) in the rounds they investigate, toxic experts can examine groupings of the model underscores from different preparation runs with various hypercutoff points, and in cross-gadget FL, harmful contraptions can have either white-box or disclosure acceptance to the final model. Thus, to give comprehensive securities not in favor of outside enemies, it is essential to consider what can be gained from the moderate repeats and final model.

4.3.1 Auditing the Iterates and Final Model

To much more rapidly understand what can be acquired from the center rehashes or final model, we propose assessing joined learning model deficiency toward specific attacks. This is a particularly captivating issue concerning the joined understanding the setting. As indicated by one point of view, adversaries can get quick enrollment to the model from the specialist, which develops the attack surface. Clearly, the master figures out in which specific seasons of the blueprint cooperation the enemy will receive an acknowledgment to the model, and also controls the adversary's outrageous influence at all the stages.

For excellent (nonconsolidated) models of appraisal, understanding a model's inadequacy to attacks is a working and testing research area. The most striking framework for evaluating a model's weakness to an attack is to mirror the attack on the model using a middle individual (investigating) dataset such as the dataset expected ultimately. This plan allows considering what the model's standard attack insufficiency is if the middle individual dataset is in fact similar to the inevitable customer data. A safer strategy is to pick the most adverse condition upper bound on the model's attack weakness. This condition can be drawn nearer hypothetically, yet this regularly yields free, vacuous cutoff focuses for sensible models. Cautious frameworks may have the decision to give even more close cutoff focuses, yet for such assaults and models, this undertaking might be unmanageable. A captivating arising zone of examination regarding this space takes apart the speculative conditions (on the surveyed model and assaults) under which an inefficient endeavor to see security infringement by a duplicated assault suggests that no more grounded assaults can succeed at such an undertaking. In any case, this district is now beginning and more work should never really be able to acquire the key prerequisites under which surveying (by strategies for reenacted assaults) is sufficient.

The joined learning structure gives an interesting setting for attacks, yet alongside attack quantification and protection. Specifically, in view of the specialist's association over when each customer can access and influence the model during the organizing cycle, it very well may be achievable to design new sensible techniques for assessing a model's basic case or most suspicious situation attack shortcoming. Such strategies would associate with the advancement of new adaptable watchmen, which can be applied on the fly to obtain a significant genuinely arranged influence while boosting utility.

4.3.2 Training with Central Differential Privacy

To limit or execute the data., that could be several courses of action concerning a person from the underlines (correspondingly as final model), client level differential security can be utilized in the FL iterative arranging measure [7, 29]. With this procedure, the expert cuts the two rules of individual updates, amounts to the cut updates, and thusly adds Gaussian commotion to the total. This step guarantees that the enhancement does not overfit to any individual client's update. To follow the general affirmation money-related plan across changes, progressed association hypotheses or the authentic second agent framework made can be utilized. The second's agent system works fittingly with the dependably subsampled Gaussian instrument.

In cross-contraption FL, the extent of preparing models can change clearly beginning with one gadget then onto the going with. Thusly, for example, advancing abroad client-level DP in the focal model [24], figuring out how to adaptively bound the commitments of clients and catch very far overabundance parts a mind-boggling appraisal bearing. Essentially more all around, not in any manner like.

To mask the rehashes from the clients, each client could run their nearby section of joined learning inside a TEE giving confidentiality features. The specialist would accept that the normal joined learning code is running in the TEE (subject to the TEE's confirmation and immovable quality features), then give an encoded model rehash to the device with an authoritative objective that it should be unscrambled inside the TEE. Finally, the model updates would be encoded inside the TEE before being returned to the prepared proficient, using keys only known inside the zone and on the specialist. Incomprehensibly, TEEs may not be all through open-across clients, especially when those clients are end-customer contraptions, for instance, cells. Additionally, regardless, when TEEs are free, they may not be sufficiently famous to assist with masterminding appraisals, which would have to happen inside the TEE to guarantee the model rehash, and conceivably computationally ridiculous correspondingly as require significant degrees of RAM–regardless, TEE limits are without a doubt going to improve as time goes on, and procedures, for instance, those presented here, have the choice to decrease the fundamentals on the TEE by exchanging pieces of the figuring outside the TEE while keeping up the affirmation, ordinariness, and confidentiality needs of the count.

A near certificate can be created under the MPC model [14]. For example, the master could scramble the underline's model cutoff focuses under a homomorphic encryption plot before sending it to the client, using keys known particularly to the specialist. The client could then enroll the mixed model update using the homomorphic properties of the cryptosystem, without hoping to interpret beyond what many would consider possible. The mixed model update could then be returned to the laborer for assortment. A fundamental test here will be to drive a combination on the master before unscrambling, as some different option from what is generally anticipated, the laborer may have the choice to get settled with a client's model update. Another awkward open issue here is improving execution, as bleeding-edge systems can require significant computational resources to complete a single round of getting ready in a basic neural alliance. Progress here could be made both by algorithmic advances equivalently as through the improvement of more efficient gear reviving specialists for MPC.

Additional troubles arise if the model rehashes should in like manner be hidden from the specialist. Under the TEE model, the masterpiece of joined learning could run inside a TEE, with every get-together (i.e., clients and prepared experts) watching that the laborer TEE will simply ship the final model after the appropriate organizing rules have been met. Under the MPC model, an encryption key could obtain the model highlights, with the key held by the analyst, dispersed in parts between the clients, or held by a trusted source at a distance; in this plan, the keyholder(s) would be depended upon to investigate the loosening of distant, and could along these lines ensure that this cycle happens only in solitary time.

4.3.3 Repeated Analyses Over Evolving Data

For explicit employments of joined learning, the master wishes to research data that appear in a streaming game plan, and furthermore should provide honestly revived learned models that (1) are directly on the data seen up until this point, and (2) precisely anticipate future data appearances. Without security concerns, the inspector could fundamentally retrain the learned model once new data appears, to ensure the main precision reliably. Regardless, because security guarantees are horrible as additional information is spread about equivalent data, these updates ought to be less unsurprising to save both the certification and exactness of the overall evaluation.

Progressing prompts in differential security for dynamic informational indexes and time-methodology data have all the ordinary presence of a trusted controller who can acknowledge unsavory to be as they appear on the web, and pass on suitably fortified snippets of data. Open sale is a methodology by which these algorithmic procedures can be loosely dependent upon the set setting, to associate with private joined learning on-time outline data or other genuine driving enlightening combinations.

4.3.4 Preventing Model Theft and Misuse

Sometimes, the entertainer or organization building up an ML model might be persuaded to bind the capacity to research, abuse, or take the model. For instance, confining consent to a distance may make it more difficult for an enemy to look.

4.4 Challenges: Communication Channels, Sybil Attacks, and Selection

In the cross-contraption FL setting, we have a specialist with significant computational resources and inestimable clients that (i) can fundamentally chat with the laborer (as in a star network geography), and (ii) may be limited in accessibility and move speed. This position remarkably strong necessities while preferring a given trust model. In particular, clients do not have a sensible technique for opening up secure channels among themselves or the laborer. This caveat suggests, for reliable settings, that needing to be prompt (or conceivably semi-genuine) lead by the master in a key dispersal stage is required in conditions where private channels among clients are required. This plan combines cryptographic plans subject to MPC approaches. A decision instead of this weakness would combine an extra gathering or a public vehicle board into the model that is known to the customers and trusted to not intrigue the trained professional. Past enduring the master to help private correspondence channels, individuals in cross-contraption FL should in like path trust in the laborer to chart accomplices of clients in a sensible and genuine manner [33]. An adequately poisonous foe controlling the specialist could reenact innumerable fake client devices or could particularly pick really subverted contraptions from the pool of open devices. Regardless, the adversary could oversee evidently a more obvious number of individuals in the development of FL than would be average just from a base speed of repudiating devices in everyone. This view makes it fourfold more straightforward to break the ordinary vulnerability in MPC that on any occasion a particular piece of the contraptions is prompted, likewise sabotaging the security of the show. Whether fundamentally the security of show stays ideal (for example, if its security is set up in a substitute wellspring of trust, similar to an ensured region), there is a risk that if interminable truly engineered clients' model updates are known to or obliged by the foe, then the security of the extra clients' updates may be sabotaged. Note that these concerns can furthermore apply to TEEs. For example, a TEE-based shuffler can likewise be desperate upon a Sybil attack; if a specific reasonable customer's data are shuffled with known obligations from fake customers, it will be immediate for the enemy to see the authentic customer's motivation in the shuffled yield.

Note that every so often it may be possible to make attestation among the clients in around that they are throughout executing the correct show, for instance, if secure regions are open on client devices and the clients can remotely embrace one another. In these cases, it very well may be achievable to set up certification for all real individuals in the round (e.g., by standing up to stresses that guarantee multi-party estimation shows were followed surely, that appropriated differential security commitments were added unassumingly and adequately, etc.), whether or not simply the model updates are known to or obliged by the foe.

4.5 Limitations of Existing Solutions

Given that the objective of FL is for the worker to create a model of everyone-level models in the customers' information, a brand name security objective is to evaluate, and provably limit, the expert's capacity to copy an individual customer's information. This step joins officially defining (a) what is the perspective on the customers' information revealed to the expert because of a FL execution, and (b) what is the confirmation spillage of such a view. In FL, we are especially excited about ensuring that the worker can account for reports from the customers while by one way or another veiling the duties of every individual customer.

Nearby differential privacy (LDP), when actually investigated, sheds the need for an acknowledged focal worker by having every customer play out a differentially private change to their report before sending it to the focal subject matter expert. LDP expects that a client's security comes exclusively from that client's option of their own irregularity; accordingly, a client's protection ensures is free of the extra intercession joined by any abundance of clients. While LDP shows are persuading at endorsing security and have hypothetical justifications, various outcomes have shown that accomplishing nearby differential confirmation while ensuring utility is attempting security especially, particularly in high-dimensional information settings. A part of this difficulty is credited to the way that the level of the capricious rattle natural should be near with the size of the sign in the information, which may require blending reports between customers. Accordingly, attaining utility with LDP basically undefined from that in the focal setting subsequently requires a generally more prominent client base or a more noteworthy decision of ε limit.

Cross-assortment differential affirmation, the mutt model for differential security, can help decrease the size of the basic client base by conveying subjects to their trust inclinations; in any case, it does not give protection amplification of clients' covertly added information. Also, it is confusing which application areas and assessments can best use mixed trust model information. Current work on the cross-assortment model routinely recognizes that when paying little mind to the client trust propensity, their information comes from a similar dispersion [31]. Delivering this thought that is basic for FL unequivocally, as the relationship between the trust propensity and genuine client information, might not be minor.

The shuffle model empowers security amplification from clients' furtively added disturbance; at any rate, it goes with two problems of its own. The first is the fundamental of confiding in a delegate; if clients are beginning now to not trust the manager, it might be extraordinary that they will acknowledge a go-between that bore witness to or that made by the guardian (at any rate, TEEs may assist with overcoming this issue). The Prochlo structure is (to the most surprising point of our arrangement) the singular existing occasion. The subsequent obstruction is that the shuffle model's differential confirmation ensures ruins concerning the amount of not very organized clients taking an interest in the calculation. Since this number is not known to the clients or the regulator, it conveys a shortcoming to the confirmed degree of safety that clients are getting. This danger is especially enormous regarding joined learning because clients (who are conceivably contradicting) are a basic piece of the computational pipeline. The safe multi-party assessment, regardless of adding significant figuring and correspondence overhead to every client, does not address this danger when clients are adding their own disturbance locally.

This plan consolidates cryptographic blueprints subject to MPC approaches. A choice rather than this vulnerability would merge an additional party or a public transport board into the model that is known to the clients and trusted to not interest the specialist. Past tolerating the expert to help private correspondence channels, the people in cross-contraption FL should in a similar way trust in the worker to diagram partners of customers in a reasonable and true way [33]. A sufficiently toxic enemy controlling the worker could reenact countless phony customer devices or could especially pick actually sabotaged contraptions from the pool of accessible gadgets. In any case, the enemy could manage undeniably a more conspicuous number of people in a movement of FL than would be run of the mill just from a base speed of contradicting contraptions in everybody. This case would make it far simpler to break the commonplace uncertainty in MPC that in any event a specific piece of the contraptions is immediate, accordingly undermining the security of the show. Regardless of whether basically the security of show stays ideal (for instance, if its security is set up in a substitute wellspring of trust, like a guaranteed area), there is a hazard that if endless seriously masterminded customer model updates are known to or constrained by the adversary, by then the security of the additional customer updates might be undermined. Note that these worries can additionally apply to TEEs. For instance, a TEE-based shuffler can also be destitute upon a Sybil assault; if a particular sensible client's information is shuffled with known responsibilities from counterfeit clients, it will be direct for the foe to see the genuine client's inspiration in the shuffled yield.

Note that once in a while, it very well might be feasible to create affirmation among the customers that they are all around executing the right show, for example, if secure areas are open on customer devices and the customers can distantly endorse each other. In these cases, it might be feasible to set up affirmation for all genuine people in the round (e.g., by remaining worried that ensured multiparty calculation shows were followed certainly, that appropriated differential security obligations were added unpretentiously and effectively, and so on) regardless of whether just the model updates are known to or obliged by the enemy.

4.5.1 Limitations of Existing Solutions

Given that the goal of FL is for the laborer to make an everybody-level model in the client data, a brand name security objective is to assess, and provably limit, the master's ability to duplicate an individual client's data. This task joins formally defining (a) what is the point of view on the client data shown to the master by an FL execution, and (b) what is the affirmation spillage of such a view. In FL, we are particularly anxious to guarantee that the laborer can add to reports from the clients while by some means veiling the obligations of each individual client.

Nearby differential privacy (LDP), when really examined, sheds the requirement for a recognized central specialist by having each client play out a differentially private change to their report before sending it to the central topic master. LDP expects that a customer's security comes only from that customer's choice of their own singularity; in like manner, a customer's insurance guarantees is liberated from the additional intervention when joined by a bounty of customers. Although LDP shows are convincing at supporting security and have theoretical justifications, different results have shown that achieving nearby differential affirmation while guaranteeing utility is particularly challenging, especially in high-dimensional data settings. A part of this difficulty is credited to the way that the level of the fanciful clatter characteristic should be close with the size of the sign in the data, which may require mixing reports between clients. In like manner, attaining utility with LDP essentially indistinct from that in the central setting in this way requires, by and large, a more conspicuous customer base or more significant choice of ε limit.

Cross-arrangement differential insistence, the mutt model for differential security, can help decay the size of the essential customer base by passing on subjects to their trust tendencies; regardless, it does not give insurance amplification of upheaval secretly added by the customers. Plus, it is confounded which application territories and evaluations can best utilize blended trust model data. Current work on the cross-grouping model regularly perceives that paying little psyche to the customer trust inclination, their data come from a comparative scattering [31]. Conveying up this idea that is essential for FL unequivocally, as the connection between the trust inclination and certified customer data may not be minor.

The shuffle model engages security amplification from customers' quickly added unsettling influence, and at any rate it has two irritants of its own. The first is the major of a trusted delegate: if customers are starting now to not trust the supervisor, it very well may be unprecedented that they will recognize a go-between which demonstrated the veracity of or made by the gatekeeper (at any rate, TEEs may help with defeating this issue). The Prochlo structure is (to the most stunning mark of our course of action) the particular existing event. The resulting impediment is that the shuffle model differential affirmation guarantees ruin concerning the measure of not too many coordinated customers looking into the computation. Because this number is not known to the customers or the controller, it passes on a weakness to the affirmed level of security that customers are getting. This risk is particularly monstrous with respect to joined learning because customers (who are possibly negating) are a fundamental piece of the computational pipeline. The safe multi-party evaluation, paying little mind to adding significant figuring and correspondence overhead to each customer, does not address this risk when customers are adding their own unsettling influence locally.

This framework is formalized in the pufferfish improvement of well-being, which allows the investigator to show a class of guaranteed predicates that should be learned on the confirmations of differential insurance, and any additional predicates can be learned without differential security. For this approach to manage developing some way to give stunning certification guarantees considering everything, the expert should fathom the customers' security needs to their particular assessment undertaking and data blend structure. The unified learning design could be modified to allow explicit customers to figure out what surmises they allow and disallow. These data hindrances could either be set up on the device, with just "appropriate" information being surrendered to the master in the FL model update step, or could be conceivable as a piece of the assortment adventure at whatever point data have been amassed. Further work should be done to make explicit inventions for setting such customer tendencies into the FL model and to make strategies for fundamental affinity elicitation from customers.

4.5.2 Behavioral Research to Elicit Privacy Preferences

Any approach to manage or regulate security that requires explicit customers to show their own insistence rules should harden either lead or field assessment to ensure that customers can pass on educated tendencies. This effort should combine both an informative part and a tendency assessment.

The enlightening part should measure and improve customer appreciation of the security movement being used and the nuances of data use [35]. For applications including joined learning, this should in like way harden explanations of set learning and definite what data will be sent to the topic master. When the enlightening part of the assessment has verified that common customers can thoroughly fathom the security guarantees offered by a private learning measure, then specialists can begin affinity elicitation: this can happen in lead labs, goliath development field appraisals, or insignificant focus get-togethers. Care should be utilized to ensure that individuals giving data on their tendencies are both sufficiently instructed to give first in class data and are illustrative of the goal people.

Although the rich field of social and exploratory cash-related focuses has for some time shown that people act contrastingly in open versus private conditions (that is, whether their choices are seen by others or not), close to no lead work has been done on bringing out penchants for differential security [10, 12]. Developing this calling will be a focal progression toward unfathomable future utilization of private joined learning. Results from the instructive study will show significance here in guaranteeing that overviewed people are completely taught and like the choices toward which they are going. It should be a basic tenant of these examinations that they are performed morally and that no dubiousness is consolidated.

5 Force to Attacks and Failures

Current AI developments can be powerless against various kinds of disappointments. These failures join non-unsafe confusions, for instance, bugs in pre-managing pipelines, tumultuous masterminding names, interesting clients moreover as express attacks that target planning and sending pipelines. All through this section, we reliably see that the appropriated nature, building plan, and data necessities of joined learning open up new frustration modes and attack surfaces. Furthermore, security frameworks to guarantee affirmation in joined learning can make seeing and assuaging these mistakes and attacks a particularly testing task.

Although this confluence of troubles may make strength difficult to achieve, we will look at many enabling headings of study, correspondingly as how they may be acclimated to or improved in joined settings. We also discuss broad requests concerning the connection between different sorts of attacks and upset assumptions, and the significance of these relationships in joined learning.

Backing: in the proposed strategy, the aforementioned devices are used to screen a normal current workspace during human–robot collaboration (HRC) attempts to perceive and follow the circumstance of the human directors (i.e., they appear a distance away from individuals) that are moving nearby a mechanical regulator inside a fenceless space. In now-shared workspaces, assessing positions and distance is critical to keep an expert demand system, to execute crash avoidance, a decline of speed, anticipating contacts of a confined substance, etc. Also, it is basically captivating that heads are freed from wearable devices to pass on zone information [3]. Following body upgrades likewise should not depend on amazing human associations. For the static establishment, the issue of the torpid body perceiving proof and running can be tended to through establishment settlement methods and ML devices [34, 36]. The presence of the robot, dependably depicted by an enormous metallic size, that moves inside the central workspace addresses extra efforts in proceeding organizing because robots actuate tremendous, non-fixed, and vivacious RF-disrupting impacts [32].

The radars total a gigantic heap of data that cannot be conveyed back to the specialist for status and affirmation because of the slowness blocks constrained by the expert security outlines. Basically, direct correspondence with the laborer is open regardless of how saving to screen of the robot works out (and rethinking modernized tasks in case of hazardous conditions) [32] and should not be used for data scattering. As needs are, to settle the adaptability challenge while paying special mind to laziness, relentless quality, and information move limit, we let the contraptions perform the model getting ready with no master coordination other than using basically regular interest with neighbors. From now on we are embracing the proposed in-network FL figurations, relying simply on nearby model exchanges over the D2D dynamic affiliations.

5.1 Conclusions and Open Issues

This paper looked for another social occasion of FL strategies that sway the normal help of gadgets in dispersed distant IoT networks without depending upon the help of a focal facilitator. The difference in joined averaging, to be unequivocal in the CFA framework, was first talked about to mistreat passed-on game plan standards for FL. By then, to improve blend speed, we proposed another figuring subject to the iterative trade of model updates and inclinations. The CFA-GE assessment was improved for two-stage propensity exchanges so it very well may be changed to self-unequivocally gigantic expansion affiliations.

The proposed dispersed learning approach was asserted on an IIoT condition where a NN model was distributedly set up to manage the issue of aloof body affirmation inside a human–robot total workspace. CFA-GE was shown to accomplish the show of worker side (or centered) joined the movement. Decentralized streamlining is absolutely worker-less concerning NN model preparation as information is moved down into the IoT devices. This aspect is especially persuading when direct correspondence with the framework is set to the side for major undertakings (e.g., to control the robot or to perform growing undertakings). Excited by the increasing degree of ML applications that will be sent in 5G and past distant affiliations, FL by techniques for the course of action arises in this paper as a promising development for flexible model movement over affiliations portrayed by decentralized association plans as in titanic IoT executions.

Dismissing the promising highlights, the proposed understanding-based framework is offering to acclimate to the situation, which should be considered during the design plan. For instance, the paper considered a crucial enough NN model for development running on IoT contraptions with bound assessment limits. Regardless, arranging further relationships on obliged gadgets is expected to become standard sooner rather than later. The utilization of comprehension-based joined improvement to more huge affiliations may require more proficient utilization of the kept data move limit, including quantization, squeezing factor, or spontaneous channel encoding. As seen in the contemplated setting-focused assessment, tweaking of the model hyperlimits [16] correspondingly as pushing the learning rates for each NN model layer unreservedly are also sensible reactions for limiting the number of correspondence changes. Finally, the paper looked for a depiction task such as the utilization of joined redesign. Both CFA and CFA-GE can be summed up to play out a wider range of puppets.

References

1. S. Savazzi, M. Nicoli, V. Rampa, S. Kianoush. Federated learning with mutually cooperating devices: A consensus approach towards server-less model optimization. ICASSP2020, pp. 3937–3941
2. S. Savazzi et al. Federated learning with cooperating devices: A consensus approach for massive IoTNetworks. IEEE Internet Things J. (2020). https://doi.org/10.1109/JIOT.2020.2964162
3. M. HabiburRehman, K. Salah, E. Damiani, D. Svetinovic. Center for Cyber-Physical System, Towards blockchain-based reputation-aware federated learning. IEEE Explorer (2020), pp. 183–188
4. J. Konečný et al. Federated learning: Strategies for improving communication efficiency, CoRR (2016). [On-line]. Available: http://arxiv.org/abs/1610.05492
5. K. Bonawitz et al. Towards federated learning at scale: System design. Proceedings of the 2nd SysML Conference, Palo Alto, CA, USA (2019)
6. S. Wang et al., Adaptive federated learning in resource constrained edge computing systems. IEEE J. Selected Areas Commun. **37**(6), 1205–1221 (2019)
7. W.G. Hatcher et al., A survey of deep learning: Platforms, applications and emerging research trends. IEEE Access **6**, 24411–24432 (2018)
8. M. Bennis et al., Ultrareliable and low-latency wire- less communication: Tail, risk, and scale. Proc. IEEE **106**(10), 1834–1853 (2018)
9. G. Soatti et al., Distributed signal processing for dense 5G IoT platforms: Networking, synchronization, interference detection and radio sensing. Ad Hoc Network **89**, 9–21 (2019) ISSN 1570-8705
10. E. Jeong et al., *Communication-Efficient On-Device Machine Learning: Federated Distillation and Augmentation under Non-IID Private Data* (NIPS Workshop, Montreal, Canada, 2018)
11. I.D. Schizas et al., Consensus in ad hoc WSNs with noisy links–part I: Distributed estimation of deterministic signals. IEEE Trans Signal Proc. **56**(1), 350–364 (2008)
12. R. Olfati-Saber et al., Consensus and cooperation in networked multi-agent systems. Proc. IEEE **98**(7), 1354–1355 (2010)
13. C. Fantacci et al. Consensus labeled random finite set filtering for distributed multi-object tracking [Online]. Available: https://arxiv.org/abs/1501.01579
14. G. Soatti et al., Consensus-based algorithms for distributed network state estimation and localization. IEEE Trans. Signal Inf. Proces. Over Netw. **3**(2), 430–444 (2017)
15. S. Samarakoon et al. Federated learning for ultra-reliable low-latency V2V communication. Proceedings of the IEEE Globecom (Dec. 2018)
16. M.G. Rabbat et al., Quantized incremental algorithms for distributed optimization. IEEE J. Selected Areas Commun. **23**(4), 798–808 (2005)
17. F.S. Cattivelli et al., Analysis of spatial and incremental LMS processing for distributed estimation. IEEE Trans. Signal Proc. **59**(4), 1465–1480 (2011)
18. L. Bottou, Large-scale machine learning with stochastic gradient descent. Proceedings of COMP-STAT, Physica-Verlag HD (2010)
19. M. Blot et al., Gossip training for deep learning. 30th Conference on Neural Information Processing Systems (NIPS), Barcelona, Spain (2016). [Online]. Available: https://arxiv.org/abs/1611.09726
20. J.A. Daily et al., Gossipgrad. Scalable deep learning using gossip communication based asynchronous gradient descent [Online]. Available: https://arxiv.org/abs/1803.05880
21. A.H. Sayed et al., Diffusion strategies for adaptation and learning over networks: An examination of distributed strategies and network behavior. IEEE Signal Proc. Mag. 30(3) (2013)
22. N.D. Lane et al., Squeezing deep learning into mobile and embedded devices. Pervasive Comput. **16**(3), 82–88 (2017)
23. Silicon Radar, 120-GHz highly integrated IQ transceiver with antennas on chip in silicon germanium technology (Nov. 2018). Available: https://siliconradar.com/datasheets/Datasheet TRA 120 002 V0.8.pdf

24. V. Rampa et al. Safe human–robot cooperation through sensor-less radio localization. Proc. 12th IEEE INDIN, pp. 683–689 (2014)
25. S. Kianoush et al., Device-free RF human body fall detection and localization in industrial workplaces. IEEE Internet Things J. **4**(2), 351–362 (2017)
26. L. Ascorti et al., A wireless cloud network platform for industrial process automation: Critical data publishing and distributed sensing. IEEE Trans. Instr. Meas. **66**(4), 592–603 (2017)
27. P. Kairouz, H. Brendan McMahan, B. Avent, A. Bellet, M. Bennis, A.N. Bhagoji, K. Bonawitz, Z. Charles, G. Cormode, R. Cummings, G.L. D'Oliveira Rafael, S. El Rouayheb, D. Evans, J. Gardner, Z. Garrett, A. Gascón, B. Ghazi, P. B. Gibbons., M. Gruteser, Z. Harchaoui, C. He, L. He, Z. Huo, B. Hutchinson, J. Hsu, M. Jaggi, T. Javidi, G. Joshi, M. Khodak et al., Advances and open problems in federated learning (10 December 2019). arXiv:1912.04977 [cs.LG]
28. S.R. Pokhrel, Federated learning meets blockchain at 6G edge: A drone-assisted networking for disaster response, 49–54 (2020). https://doi.org/10.1145/3414045.3415949. S2CID 222179104
29. A.M. Elbir, S. Coleri, Federated learning for vehicular networks (2 June 2020). arXiv:2006.01412 [eess.SP]
30. M.G. Arivazhagan, V. Aggarwal, A.K. Singh, S. Choudhary, Federated learning with personalization layers (2 December 2019). arXiv:1912.00818[cs.LG]
31. M.M. Amiri, D. Gunduz, Federated learning over wireless fading channels (10 February 2020). arXiv:1907.09769 [cs.IT]
32. H. Brendan McMahan, D. Ramage, P. Richtárik, Federated optimization: Distributed machine learning for on-device intelligence. Jakub Konečný (2016).
33. X. Xian, X. Wang, J. Ding, R. Ghanadan, Assisted learning: A framework for multi-organization learning. Adv. Neural Inf. Proc. Syst. 33 (2020). arXiv:2004.00566
34. R. Cioffi, M. Travaglioni, G. Piscitelli, A. Petrillo, F. De Felice, Artificial intelligence and machine learning applications in smart production: Progress, trends, and directions. Sustainability **12**(2), 492 (2019). https://doi.org/10.3390/su12020492
35. N. Rieke, J. Hancox, W. Li, F. Milletarì, H.R. Roth, S. Albarqouni, S. Bakas, M.N. Galtier, B.A. Landman, K. Maier-Hein, S. Ourselin, M. Sheller, R.M. Summers, A. Trask, D. Xu, M. Baust, M.J. Cardoso, The future of digital health with federated learning. NPJ Digital Med. **3**(1), 119. arXiv:2003.08119 (2020). https://doi.org/10.1038/s41746-020-00323-1. PMC 7490367. PMID 33015372. S2CID 212747909
36. B. McMahan et al., Communication-efficient learning of deep networks from decentralized data. Proc. 20th Int. Conf. Artificial Intel. Stat **54**, 1273–1282 (2017)

A Prospective Study of Federated Machine Learning in Medical Science

Rijwan Khan, Mahima Gupta, Pallavi Kumari, and Narendra Kumar

1 Introduction

The wide-ranging effect of healthcare on billions of individuals worldwide has urged tremendous interdisciplinary exploration endeavours and momentous developments. The chance of day-to-day human services monitoring, related to ordinary clinical medicinal services, vows to introduce another era of smart healthcare. Smart healthcare services utilize the new epoch of data innovations, for example, the internet of things (loT), machine learning, big data, natural language processing (NLP) and artificial intelligence, to change the customary clinical framework in an overall manner, making human services more proficient, helpful and customized. At present, medicinal services improve personal satisfaction.

R. Khan (✉)
Department of Computer Science and Engineering, ABES Institute of Technology, Ghaziabad, UP, India

M. Gupta
Scholar, Department of Computer Science and Engineering, ABES Institute of Technology, Ghaziabad, UP, India

P. Kumari
Scholar, Department of Computer Science and Engineering, ABES Institute of Technology, Ghaziabad, UP, India

AKTU Lucknow, Lucknow, UP, India

N. Kumar
Dean Collaborations and Partnerships Bluecrest University College, Monrovia, Liberia
e-mail: narendra.kumar@bluecrestcollege.com

S. P. Yadav et al. (eds.), *Federated Learning for IoT Applications*, EAI/Springer Innovations in Communication and Computing,
https://doi.org/10.1007/978-3-030-85559-8_7

Artificial intelligence (AI) is the term used to delineate the use of PCs and development to derive direct and essential conclusions for all intents and purposes indistinguishable from a person. John McCarthy initially portrayed the term AI in 1956 as the science and building of making smart machines.

Artificial intelligence in medicine and human services has been a particularly fervently discussed issue recently. While there is a tendency of extraordinary potential in the use of AI in medicine, there are additional worries over the loss of the 'human touch' in such an essential and individual focused calling. Re-enacted insight in drugs suggests the use of AI advancement/robotized structures in the diagnosis and treatment of patients who require care [2, 7]. While determination and treatment may appear straightforward advances, numerous other foundation forms must occur for a patient to be appropriately dealt with, for instance:

- Deciding a proper treatment strategy (frequently introducing alternatives).
- Preparing and directing the picked treatment strategy.
- Patient monitoring.
- Aftercare, follow-up arrangements, and so on.

The potential for extended AI use in medication is not just in a decline of manual assignments and the opening up of specialist's time, growing viability and proficiency; moreover, it presents the open entryway for us to move towards more 'precision medicine'. Artificial intelligence methods, for instance, modulated expert structures, Bayesian neural networks, artificial neural networks and smart security systems, have been used in different clinical settings in human services [6]. AI is becoming more commonplace in the clinical field, ranging from online enrolment in clinical centres, digitization of clinical records, update calls for follow-up courses of action and immunization rates for children and pregnant females to quiet portion estimations and opposing effect reprobation while embracing multidrug blends. PCs have been used from the outset in clinical imaging for administrative work such as image determination and now have the ability to transform into an essential part of the work environment with the wellspring of image documentation and corresponding structure. The use of computer-aided diagnosis (CAD) in screening mammography is outstanding. Continuous assessments have shown that CAD does not have a huge amount of characteristics guide, given positive judicious characteristics, affectability and particularity [12]. Radiology has been the most inviting to the utilization of progression. Artificial intelligence could give a significant guide in radiology by naming specific tests just by recognizing rapid negative tests in figured tomography, X-pillars, magnetic resonance images, especially in high volume settings and in clinical facilities with less available HR.

1.1 Artificial Intelligence In Medical And Healthcare Applications

Advances in computational power coordinated with colossal amounts of data obtained in human administrations systems have created various clinical issues for AI applications. Two computations that can benefit patients and specialists through creation assurance are briefly discussed below.

The first of these estimations is a current example of a computation that out performs specialists in image gathering assignments. In the fall of 2018, experts at Seoul National University Hospital and College of Medicine developed an AI algorithm called Deep Learning-based Automatic Detection (DLAD) for analysis of chest radiographs and to detect bizarre cell advancement, for instance, dangerous growths [1, 3]. The algorithm's introduction stood out from various specialists' known limits on comparable images and outmanoeuvred 17 of 18 specialists.

The second of these figures is from researchers at Google AI Healthcare, who made a learning computation, Lymph Node Assistant (LYNA), in the fall of 2018, that inspected histology slides recoloured tissue tests to perceive the development of metastatic chest tumours from lymph centre biopsies [4, 16]. LYNA assessed two datasets and appeared to correctly portray a model as hurtful or noncancerous precisely 99% of the time.

Regardless of the unquestionably rich AI writing in human administrations, the assessment prevalently centres on a few ailment types: harm, tangible framework disease and cardiovascular affliction.

(i). **CANCER**: Artificial intelligence (AI) approaches can possibly influence a few features of malignant growth treatment. This also incorporates medication disclosure, improvement and how these prescriptions are clinically endorsed and finally coordinated at the motivation behind consideration, among others. Currently, these systems are expensive and discouraging. Moreover, medicines as often as possible realize factor treatment results among patients [18]. The mix of AI and ailment treatment has realized various responses to address these challenges.

(ii). **DIABETES**: Diabetes is an overall pandemic. At least 425 million people worldwide have diabetes, speaking to 12% of the world's health service utilization; however, 1 out of 2 individuals remain undiagnosed and untreated. Quick advances in AI offer the assurance of making both determined and unstructured health data available for people with diabetes (PWDs). Regardless of the type of AI, social protection is represented in the composition and new AI-controlled devices are being confirmed for diabetes care, but an efficient review of clinically relevant diabetes AI applications is missing. The AI applications are intended to improve a wide range of aspects of diabetes care, from diabetes screening and discovery to checking and treatment, and include applications, devices and frameworks that guide patients, clinicians and well-being frameworks.

(iii). **NEUROLOGY:** AI has been used by radiologists as neuroimaging changed over to digitized image collections and these calculations have precision surpassing that of the prepared natural eye. Bentley et al. utilized CT filter-based AI to foresee intracranial drain in patients accepting intravenous thrombolysis with TPA. AI has likewise been utilized in result expectations and prognostic assessment after strokes.

(iv). **CARDIOLOGY:** Artificial intelligence apparatuses are demonstrating their utility in the developing field of cardiology. Notwithstanding, cardiologists must comprehend their maximum capacity to utilize them proficiently [5, 8, 9]. Nowadays, the dynamic procedure in medicine is an unpredictable assignment that, in a perfect world, depends on the accessibility of solid and target bits of proof, quick access to information, such as the best possible understanding of accessible realities with the fusion of patient advantage hazard proportions in each progression. The fusion of AI instruments in the field of cardiology into day-to-day dynamics should improve care. Obviously, in any case, cardiologists must have ultimate control of the framework, pay attention to the choices and have the power to change calculations in the cases that may turn out badly. In the field of cardiovascular imaging, AI indicates incredible potential.

Deep learning is another AI strategy, and its clinical applications have created a lot of enthusiasm recently. The across the board utilization of the internet empowers an online assortment of big data and information, and huge storage devices likewise become required. The specialized progression spurred the formation of new calculations and results that require such monstrous storage and tremendous processing abilities [10, 18]. Initially, people comprehended and controlled the practices of PCs, and now, PCs have surpassed people in memory and calculation abilities, such as the presentation of certain AI-based frameworks to tackle specific assignments.

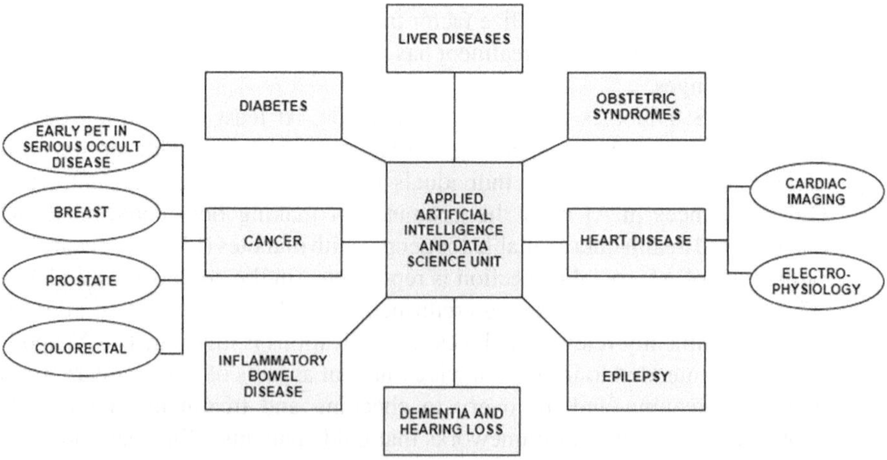

Fig. 1 Applications of artificial intelligence in disease detection

The utilization of AI has tremendous potential in clinical consideration. For example, AI-based frameworks utilizing enormous quantities of revealed analytic images previously outperformed the ability of human radiologists [17]. Be that as it may, the concern of the human–computer relationship would likewise affect clinical consideration applications. As of late, significant learning has been utilized to perceive chance factors related with cardiovascular ailments (for example, heartbeat, smoking and weight) from retinal photos. New progressions in AI restorative organizations improvement are smoothing out the patient experience, helping emergency room staff process millions, if not billions of server farms, quicker and much more profitably.

There are also an impressive number of definitive applications in social protection. The use of AI is somewhat less possibly dynamic here, but it can give liberal productivity.

1.1.1 Internet of Things in Smart Healthcare System

The internet of things is one of the foremost emerging technologies in today's world. Without the IoT, the healthcare industry faced problems such as:

- No real-time data was available.
- Lack of smart care devices.
- Inaccurate standard analytics.

There was no real-time data available; therefore researchers had to work on leftover data during clinical research. With the IoT, the devices can interact, collaborate and share experiences (by learning from each other's experiences). The IoT is a platform where regular devices are connected to the internet so that they can interact, collaborate and exchange data with each other. It reduces human intervention in

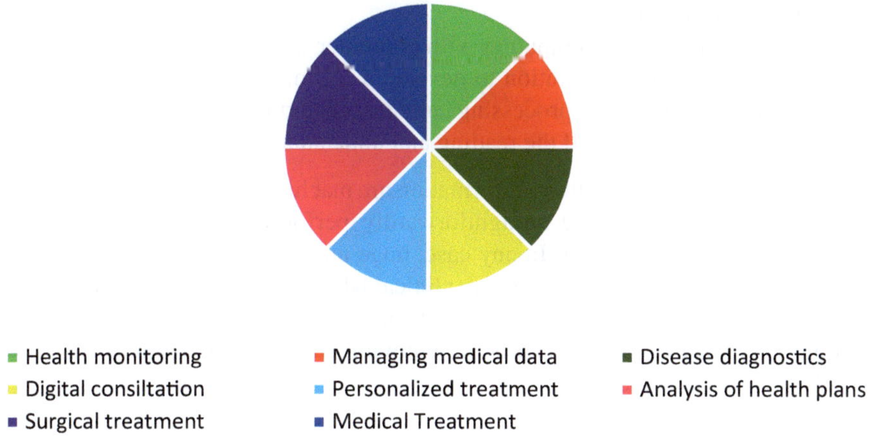

Fig. 2 Applications of artificial intelligence in healthcare

the machine cycle. The IoT has empowered the healthcare industry through the following ways:

- The IoT can provide real-time data through analysis and testing.
- The IoT improves the quality of care.
- The IoT makes devices smarter and provides far superior analytics.
- The IoT provides the advantage of remote patient monitoring by which a patient can replace the long wait in the queue by a quick check-in, data share and instruction.

1.2 Artificial Intelligence Devices

Artificial intelligence devices mostly fall into two critical classes. The chief class fuses AI machine learning (ML) methodology that separates composed data, for instance, imaging, innate and EP data. In the clinical applications, the ML procedure attempts to amass patients' characteristics or prompt the probability of the sickness results [11, 12]. The resulting class fuses natural language processing (NLP) with procedures that expel information from unstructured data, for instance, clinical notes/clinical journals to improve and progress sorted out clinical data. The NLP techniques target sending works to machine-reasonable data, which would then have the option for analysis using ML methods. As stunning as AI strategies can be, they should be spurred by clinical issues and be applied to help the final clinical practice.

1.2.1 NLP in Healthcare

- Process dictated text from larger documents to identify useful information and use it for further analysis.
- NLP can be used in combination with computer assisted coding and computer assisted clinical documentation to develop a recommendation system that suggests medical codes after processing large texts and documents instead of relying on a human to perform all the reading and understanding.

The image, EP and hereditary information are machine-reasonable, and therefore the ML calculations can be straightforwardly performed after appropriate preprocessing or quality control. In any case, huge amounts of clinical data are story text, for example, physical assessment, clinical lab reports, usable notes and release outlines, which are unstructured and unimaginable for the PC program. Under this specific circumstance, NLP focuses on removing valuable data from the story text to help the clinical dynamic.

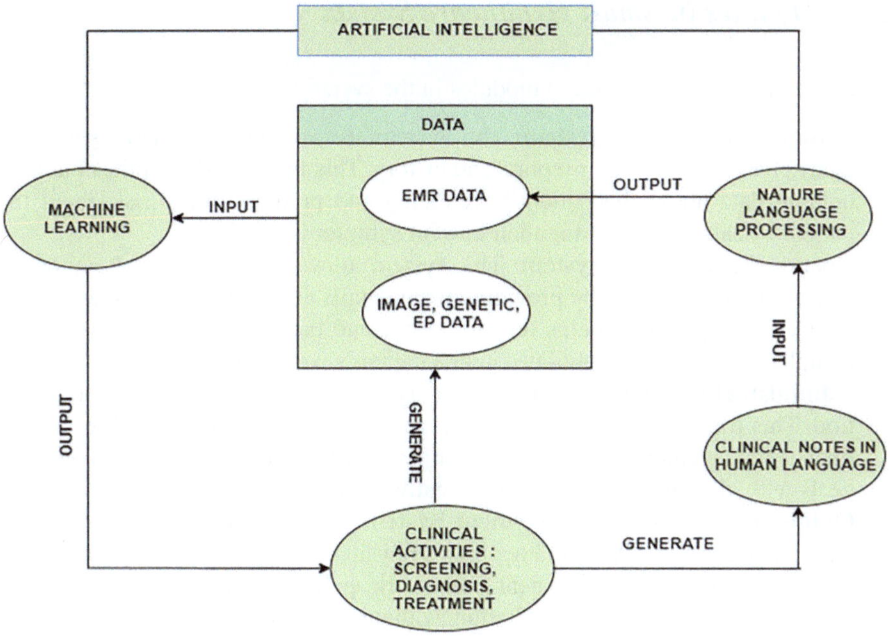

Fig. 3 The guide from clinical information age to normal language preparing information improvement, to artificial intelligence information investigation, to clinical dynamic. *EMR* electronic medical record, *EP* electrophysiological

An NLP pipeline includes two principle segments:

1. Text preparation and
2. Arrangement.

NLP can do:

- Mapping data elements present in unstructured text to structured fields.
- Conducting speech recognition to enable users to dictate notes or other information that can then be turned into text.

The combination of ailments, such as diabetes, is not thoroughly frightening. All of these diseases are factors behind death; consequently, early decisions are significant to prevent the deterioration of patients' health. Also, early discoveries can be possibly perceived through improving the assessment strategy on imaging, innate, EP or EMR, which is the nature of the AI system.

1.3 Modules in Smart Healthcare System

Smart healthcare provides these modules in the system:

1. **Patient Management System:** This system allows registered patients to enrol themselves and view their reports and history. This framework permits obtaining definite data about the patient's wellbeing and predicts future risk about the patient's health by analysing their current symptoms.
2. **Doctor Management System**: This system allows patients to make appointments with the doctors, by providing them details about their hospital as well as their clinic timings. It helps doctors to manage their duties and updates them complete arrangement subtleties with a patient's wellbeing history.
3. **Administrative Management System**: This system includes HR and organization. This framework helps in refreshing the stock record such as all records of finance management. This is a key component of organization management that deals with all purchasing and representative administration.
4. **Online Appointment Management System**: This system permits patients to have an online consultation rather than a physical visit. With the assistance of an online arrangement management framework, patients, staff and specialists can check the status of the consultation without any problem. This arrangement permits obtaining an arrangement for enlisted patients and furthermore sends updates to the client through SMS or email.
5. **Medical Services System:** This system permits including a catalogue of services that are offered by the medical clinic, for example, dental treatment administration, cardiovascular administrations, mental treatment administrations, bones authority administrations and substantially more. The patient can see the catalogue of services offered by the emergency clinic alongside every other detail of treatment. It likewise deals with the administration timing and emergency as indicated by the state of the patient.
6. **Invoice System:** After the consultation affirmation, the invoice system produces a programmed receipt against that particular patient, this receipt assists with considering the current status of instalment as well as complete instalment.
7. **Virtual Assistant:** Virtual assistants are AI and knowledge-based frameworks that associate with people to perform different tasks. This utilizes technologies such machine learning and natural language processing to empower intelligent correspondence with the end-clients. Virtual assistants in the medical services industry can help with changing different health forms and improve social insurance conveyance worldwide. It helps in improving human services quality, quiet consideration and patient results at lower costs.

2 Artificial Intelligence Medical Systems

AI-based computer-aided diagnosis (CAD) and surgical robots present a level of self-sufficiency to help in determining symptoms and treatment.

2.1 Characteristics of Artificial Intelligence Medical Systems

The artificial intelligence clinical frameworks are not the same as traditional clinical devices and frameworks according to the following three perspectives:

(i) Flexibility

The artificial intelligence clinical frameworks are capable of self-changing their exhibition through learning (i.e. adaptability). Initially, people comprehended and controlled the practices of PCs, but now PCs have surpassed people in memory and calculation capacities such as the exhibition of certain AI-based frameworks to explain specific assignments [13]. Although computer-based intelligence in clinical gadgets and frameworks with adaptability previously existed, for example, foundation commotion setting of the electro heart observation of programmed outer defibrillators, the AI frameworks can have subjectively unique adaptability from conventional devices and systems.

(ii) Predictability

Another factor influencing the impact of AI is the unpredictability of the AI yield because of the discovery idea of the AI calculations ordinarily found in deep learning. Its yield can be hard to foresee or decipher in a deductive way. This factor affects the hazard management led by the producer. The yields from AI clinical frameworks are arranged into three examples according to the consistency of the yield.

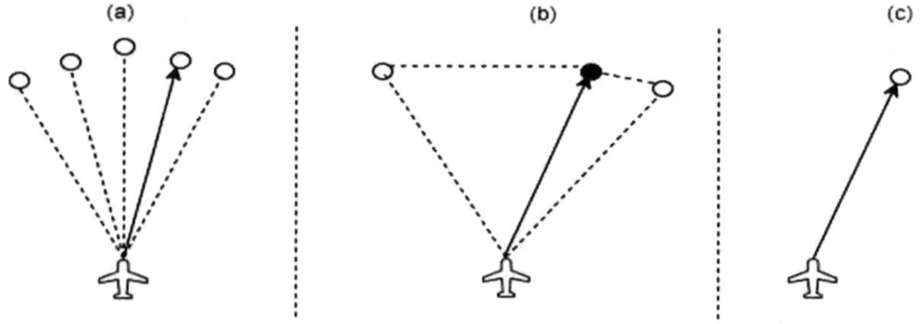

Fig. 4 Classification of artificial intelligence medical systems in terms of predictability of the output

1. Output is chosen from a guaranteed, limited number of arrangements. For instance, an AI clinical framework yields achievable finding from among foreordained choices.
2. Output is chosen from a guaranteed, limited scope of arrangements. For instance, an AI clinical framework for computer-aided analysis yields conceivable sore sites and likelihood in a scope of 0 to 1.
3. Output is not foreordained. For instance, another classification of illness is made by an AI clinical framework.

(iii) Degree of Autonomy

Further development in the AI clinical frameworks may adjust the connection among specialists and patients. Independence, which is another trait of the AI clinical frameworks, has been examined from different perspectives [1, 8]. The movement of AI-based advances and a resultant further degree of autonomy will change the association between the experts and the patients. Eventually, AI may be capable of investigation or even providing treatment plans without human intercession.

3 Types of Artificial Intelligence Medical Systems

3.1 Artificial Intelligence CAD

Artificial intelligence CAD is a run of the mill case of AI-based symptomatic frameworks. The principle contribution to the AI CAD is the symptomatic images. The AI that the AI relies upon is currently present in a specific way in the CAD frameworks that incorporate topology enhancement and generative plan abilities [15]. 'The essential elements of these highlights inside CAD is to robotize the expository strides of plan'. The key focal point of AI in CAD right now is plan enhancement that is accomplished through the development of increasingly smart structures that are lighter, more grounded and progressively affordable. The AI CAD can be ordered into five demonstrative help levels (DL1 to DL5) as follows.

DL1: AI CAD computes and presents the component identified with an illness (for example, sign of an injury, distance across or volume of a tumour) to specialists to help in the diagnosis. It is known as computer-aided diagnosis (CAD).

DL2: AI CAD figures and presents a symptomatic recommendation (for example, danger, organizing) to specialists to help in the diagnosis or to forestall oversights.

DL3: AI CAD forms images and other clinical data (multimodal data) and presents a thorough determination to specialists to help in the diagnosis.

DL4: AI CAD forms multimodal data and gives a mechanized determination to specialists. The specialists must survey and endorse the analysis.

DL5: AI CAD forms multimodal data and gives a completely robotized conclusion without doctor's contemplation and survey. It should be noted that DL1 and DL2 are found in the regular CAD used to help radiologists to forestall oversight, while DL3 and higher focus on extensive or computerized findings that may assume the job generally considered as that of specialists.

3.2 Artificial Intelligence Surgical Robot

The surgical robot with AI-based therapy assistance (AI surgical robot) is an example of an AI-based treatment system [14]. Computer-based intelligence therapeutics systems can be depicted by the quick effect of respecting the patients: AI can be related with care more clearly than by virtue of investigation. The consistency of yields and the time frame to alleviate wrong conduct by the framework likewise should be thought of.

4 Conclusion

In this chapter, owing to the development of the internet of things and artificial intelligence, we characterized a standard structure for smart human services that can advance the growth of the present-day healthcare services which are provided by most of the healthcare institutions. The IoT advances healthcare by providing real-time analysis of patients' health and predicts treatment according to symptoms. It also reduces cost and time by eliminating the wait in a long queue in the hospital for appointments. There is a wide scope of AI methods available which are prepared to deal with a variety of clinical issues. Regardless, paying little mind to earlier certainty, clinical AI advancement has not received widespread enthusiasm. The ascent of AI as a means for better human administrations offers exceptional opportunities to improve understanding and clinical gathering results, decrease costs and impact people's wellbeing. AI and natural language processing (NLP) are two critical components of artificial intelligence devices. For the ML process, the two most recognized standard procedures are neural framework and SVM. A normal AI system must have the ML part that can help to drive the sort out data, for instance, EP data, images and genetic data, and another natural language processing (NLP) module for the determination of unstructured works. Despite the unpredictability of AI information, at present, dynamic advances are happening in the subfield of neural frameworks. This has made an impact in a couple of fields of social protection science, such as prescription assessment and general wellbeing. Consequently, the goal is to improve health and create awareness among the general public and create a healthier environment.

References

1. F. Jiang et al., Artificial intelligence in healthcare: Past, present and future. Stroke Vascul. Neurol. **2**(4), 230–243 (2017)
2. K.-H. Yu, A.L. Beam, I.S. Kohane, Artificial intelligence in healthcare. Nature Biomed. Eng. **2**(10), 719–731 (2018)
3. B. McCall, COVID-19 and artificial intelligence: Protecting health-care workers and curbing the spread. The Lancet Digital Health **2**(4), e166–e167 (2020)
4. S. Reddy, J. Fox, M.P. Purohit, Artificial intelligence-enabled healthcare delivery. J. R. Soc. Med. **112**(1), 22–28 (2019)
5. J.R. England, P.M. Cheng, Artificial intelligence for medical image analysis: A guide for authors and reviewers. Am. J. Roentgenol. **212**(3), 513–519 (2019)
6. E.J. Emanuel, R.M. Wachter, Artificial intelligence in healthcare: Will the value match the hype? JAMA **321**(23), 2281–2282 (2019)
7. P. Hamet, J. Tremblay, Artificial intelligence in medicine. Metabolism **69**, S36–S40 (2017)
8. S. Kulkarni et al., Artificial intelligence in medicine: Where are we now? Acad. Radiol. **27**(1), 62–70 (2020)
9. C.P. Langlotz et al., A roadmap for foundational research on artificial intelligence in medical imaging: From the 2018 NIH/RSNA/ACR/the academy workshop. Radiology **291**(3), 781–791 (2019)
10. S.P. Yadav, K.K. Agrawal, B.S. Bhati, et al., Blockchain-based cryptocurrency regulation: an overview. Comput. Econ. (2020). https://doi.org/10.1007/s10614-020-10050-0
11. D.D. Miller, E.W. Brown, Artificial intelligence in medical practice: The question to the answer? Am. J. Med. **131**(2), 129–133 (2018)
12. W. Diprose, N. Buist, Artificial intelligence in medicine: Humans need not apply? The New Zealand Med. J. (Online) **129.1434**, 73 (2016)
13. X.-L. Du, W.-B. Li, H. Bo-Jie, Application of artificial intelligence in ophthalmology. Int. J. Ophthalmol. **11**(9), 1555 (2018)
14. G.S. Collins, K.G.M. Moons, Reporting of artificial intelligence prediction models. Lancet **393**(10181), 1577–1579 (2019)
15. D.A. Hashimoto et al., Artificial intelligence in surgery: Promises and perils. Ann. Surg. **268**(1), 70 (2018)
16. S.P. Yadav, D.P. Mahato, N.T.D. Linh, *Distributed Artificial Intelligence: A Modern Approach*, 1st edn. (CRC Press, 2020) https://doi.org/10.1201/9781003038467
17. Bhati, B. S., & Rai, C. S. (2020). Analysis of support vector machine-based intrusion detection techniques. Arabian J. Sci. Eng., 45(4), 2371-2383. Springer. https://doi.org/10.1007/s13369-019-03970-z
18. A.H. Sodhro, S. Pirbhulal, V.H.C. De Albuquerque, Artificial intelligence-driven mechanism for edge computing-based industrial applications. IEEE Transact. Indust. Informat. **15**(7), 4235–4243 (2019)

Communication-Efficient Federated Learning in Wireless-Edge Architecture

Sugandh Gupta and Sapna Katiyar

1 Introduction

Over the past few years, the field of wireless communication has emerged at a fast pace, because of the increased demand for mobile Internet and wireless empowered applications. Wireless data services were enabled by Third Generation (3G) technology for the purpose of texting and elementary web browsing, while mobile video flooding was popularized by Fourth Generation (4G) systems, and the latest 5G systems are mainly developed to support augmented reality/virtual reality, enormous levels of Internet of things (IoT), and autonomous vehicles. As of now, 5G systems have been actively deployed on a worldwide level, and researchers have started exploring the next-generation wireless technology which is the Sixth Generation (6G).

6G technology is expected to make use of huge pioneering applications and smart services by timely execution and well-organized data collection, conveyance, and manufacturing everywhere at any instant of time. Particularly, 6G will emerge as an innovative dream of ubiquitous artificial intelligence, a super elastic architecture which is capable of carrying human intelligence in almost every phase of networking system.

Artificial intelligence has been successfully applied and proven effective in various applications of our daily life. Some of the examples of AI are smart home appliances, smart vehicles, speech recognition, natural language processing, machine vision, face recognition, computer vision, etc. The perspective of artificial intelligence in optimization and improvement of wireless networking structures is very well identified and accepted.

S. Gupta (✉) · S. Katiyar
ABES Institute of Technology, Ghaziabad, India
e-mail: sugandh.gupta@abesit.in; sapna.katiyar@abesit.in

S. P. Yadav et al. (eds.), *Federated Learning for IoT Applications*, EAI/Springer Innovations in Communication and Computing,
https://doi.org/10.1007/978-3-030-85559-8_8

Various challenges have been observed for designing an AI-enabled architecture, but one key challenge is to determine the technique via which implementation of data processing and learning of distributed data across a massive number of diverse devices can be done. Federated learning (FL) is a developing distributed artificial intelligence solution which enables data-driven artificial intelligence (AI) and machine learning (ML) on a huge capacity of distributed data that exist on mobile devices. Federated learning has the capability to execute classical training and education on diverse and hypothetically enormous networks by putting whole data confined. However, it is very early to state that federated learning is going to be recognized as one of the encouraging solutions which can successfully accomplish 6G's vision of ubiquitous AI.

This chapter introduces the perspective of federated learning in the direction of ubiquitous AI for 6G. Authors have given an overview of federated learning and its categories and performance indicators. We then gave the road map to 6G and convergence of Ubiquitous AI, federated learning, and 6G. Other sections describe the main challenges and future scope of technologies.

2 Ubiquitous Artificial Intelligence

All humans imagine a world where devices, machines, and other accessories would be very intelligent enough to simplify and elevate day-to-day life. They must be capable of acting intelligently in real-time scenario with access to the environment and providing solutions to even complex problems. Artificial intelligence is the key factor for driving revolution which is accelerating at a very fast pace. AI enables machine or devices to learn from their experiences and adjust them according to new inputs to execute as human type tasks. Therefore, in today's world, the magic of artificial intelligence is everywhere. Some commonly used areas are:

- Home automation
- Computer games
- Advanced Driver-Assistance Systems (ADAS)
- Route finding in maps
- Mobile communication
- Facebook and in other social media
- Amazon's recommendations
- Trip advisor recommendation, etc.

Machine learning is the subset of AI. To make our devices intelligent, many machine learning-based solutions are required to be integrated with devices. This is known as inference, and it gives several advantages like increased reliability, improved response time, more security with increased privacy, and effective use of network bandwidth. Therefore, power efficiency is the vital requirement to scale artificial intelligence.

If we talk about wireless communication, it becomes efficient if we integrate AI with 6G. Efficiency can be observed in terms of longer battery life, improved user experience, better security, enhanced system performance, etc. Thus it can be assumed as artificial intelligence is available ubiquitously, we may not be aware of its existence. As a tool, AI is becoming very valuable and ubiquitous that it can be treated as a sixth sense. For future, it can be forseen as Artificial intelligence is ubiquitous.

3 Federated Learning

Nowadays, mobile, embedded, and sensors-based IoT devices are major sources of data. Because of frequent use due to Internet connectivity and availability on hand, mobile devices are the primary major sources of data generation. Two versions related to data are divided between good and bad: good one is that data is available everywhere, while the bad one is, data is inaccessible. Traditional machine learning is a process where data is collected from different sources and stored in a data center. When all data is available at center, a single machine learning model is trained. This process is known as centralized learning because the data is moved from the users' devices to a data center.

The field of machine learning is constantly evolving, and its new paradigm is known as federated learning. Federated learning is a modified approach where centralized data collection and model training is not required. It trains multiple machines learning models and thereafter combines results of all models into a single unit which is available at the server.

Benefits of Federated Learning Federated Learning has many benefits in comparison to centralized machine learning:

- It ensures privacy because data remains on the user's device.
- It empowers devices for collaborative learning via which prediction model can be developed by keeping the training data undamaged on device.
- Low round trip time because updated model helps for predictions on user's device.
- Power consumption is less because models are trained on user's device.

3.1 Categories of Federated Learning

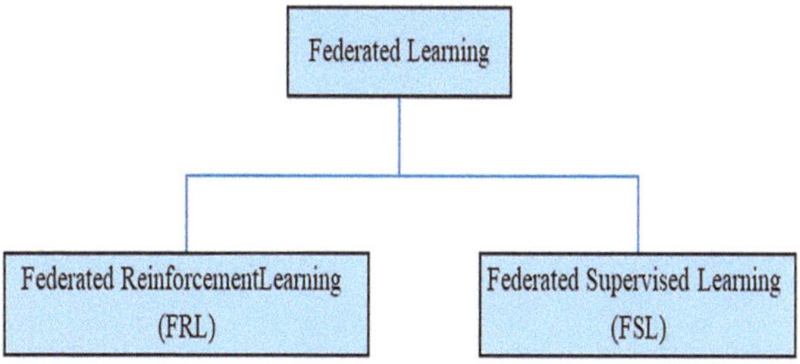

Federated Reinforcement Learning (FRL) For wireless devices, the objective of FRL is to allow devices to recall what they have learned and what other devices have learned. Each wireless device is capable of building a learning network with the help of other devices. FRL can be successfully employed when multiple wireless devices create decision in different environments.

Federated Supervised Learning (FSL) A uniform learning model is created by FSL via iterations. In each iteration, information is updated between device and base station by labelling local private data. Overall FSL procedure is executed in three steps:

- Local computation at every wireless device
- Transmission of parameters from each wireless device
- Result aggregation and broadcast at base station

3.2 Federated Learning Performance Indicators

Four main performance indicators are there.

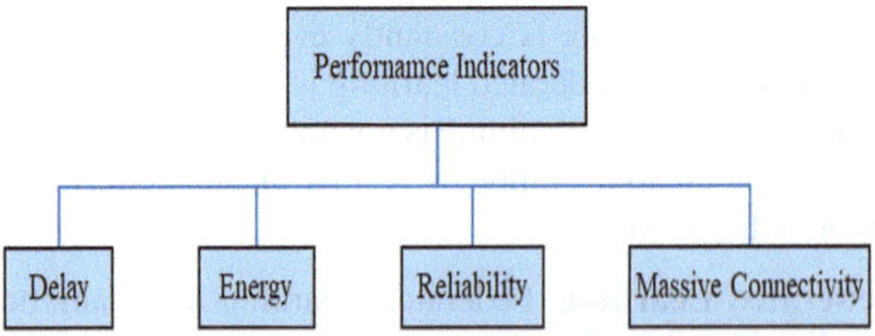

3.2.1 Delay

Federated learning delay includes:

- Local computation delay of wireless devices
- Uplink transmission delay
- Base station aggregation delay
- Downlink transmission delay

3.2.2 Energy

Energy budget is very limited for wireless devices; therefore, local computation energy (at each device, no. of iterations) and transmission energy (no. of iterations for algorithm to converge) must be measured during process.

3.2.3 Reliability

Federated learning algorithms must be trained in a distributed manner where the device communicates with the training parameters over wireless links; this is done due to limited bandwidth and unreliability of wireless links which introduces training errors.

3.2.4 Massive Connectivity

Conventional techniques of interference avoiding channel access look infeasible because they introduce excessive latency. This challenge can be overcome in multiple access channels by air computation for speedy wireless data aggregation.

3.3 Motivating Factors of Federated Learning for 6G

3.3.1 Resource Management

Sometimes nonconvex resource allocation problems arise in a multi-cell network which can be simplified by conventional algorithms like successive convex approximations, etc. Hence, there is a requirement to innovate federated learning technique which can address various resource management challenges like dynamic user clustering, joint user association, and distributed power control.

3.3.2 Communication Efficient Artificial Intelligence

By using artificial intelligence, transmitted traffic in a network can be reduced because raw data for individual devices need not be uploaded on a centralized location. Communication overhead of federated learning is an important factor which can be suppressed by reducing model update frequency and compressing data size related to individual device side.

3.3.3 User Behavior Predictions

To implement a wireless network, user behavior prediction is very important because of diverse quality of services required by users. Federated learning can be used to predict users' behavior where local FL algorithm is performed by an individual user to train a learning model; then the trained model is updated to the base station. Unified federated learning model parameters are generated by base station and transmitted to all users. Then user can dynamically select the subchannel for data uploading depending upon mobility predictions. Dynamic allocation of downlink is also done, and therefore full duplex mode of operation is performed.

3.3.4 Signal Detection and Channel Estimation

In case of wireless networks, random features of wireless channels create a key challenge for signal detection and channel estimation. Federated learning algorithm works well for downlink systems, while for multi-cell uplink systems, multi-user signals can be detected by transmitting federated learning model parameters from all base stations to server. Then unified federated learning model parameters are broadcasted from server to all base station.

3.3.5 Flexible Network Architecture

Federated learning approach has given visible and significant solutions which can be further supplemented by modern enlargements of integration of AI solutions. Some examples are generative learning, reinforcement learning, transfer learning, deep learning, etc.

Process of Federated Learning Federated learning is a very simple approach which can be summarized as follows:

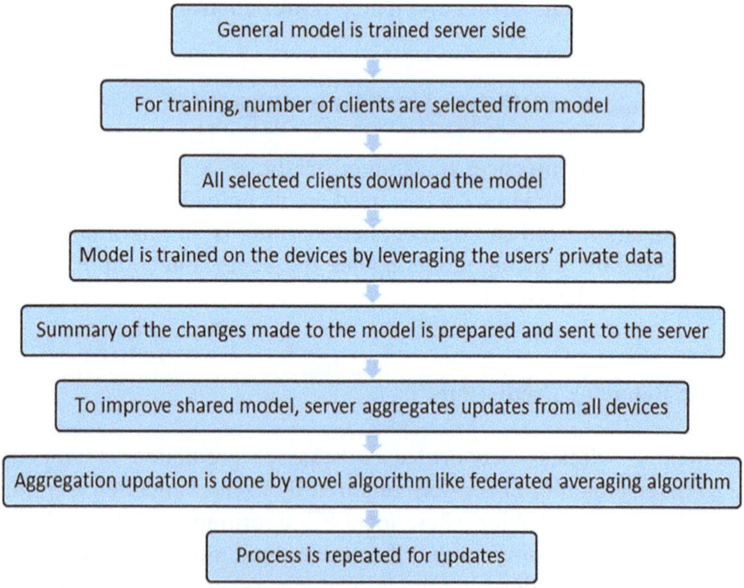

4 Road Map to 6G

The increased demands of the users of mobile radio communication have led to its evolution from Zero Generation (0G) to First Generation (1G) where analog signals were transmitted, followed by Second Generation (2G) which transmits digital signals; then came Third Generation (3G) which includes transmission at high data rate. Fourth Generation (4G) is also known as Long-Term Evolution (LTE)-supported services, which includes mobile applications, entertainment and multimedia, etc. Fifth Generation (5G) technology is developed on the approach called Open Transport Protocol (OTP) and Open Wireless Architecture (OWA). Since 5G cellular systems are being aggressively positioned everywhere now, researchers are in the progress of discovering innovative technological developments for the next generation, i.e., 6G. Alike thoughts have been given by many researchers about the integration of ubiquitous AI with 6G. It would be very elastic architecture which can carry human like intelligence in almost every to every phase of networking systems. At the same time, it was also assessed that many challenges may arise in ubiquitous AI-based Sixth Generation (6G) proposed technique (Fig. 1).

4.1 Zero Generation

Moving from wired telephone to wireless communication was a big revolution and was called Mobile Radio Telephone and when next wireless generation got created, it was called 0G or pre-cell phone mobile technology. Pioneers for this big step forward were Motorola and Bell Systems that raised the market of 0G. The technology which was used in 0G was Push-to-talk (PTT) where you press the button to speak and leave the button to listen to the other side, Mobile Telephone Service

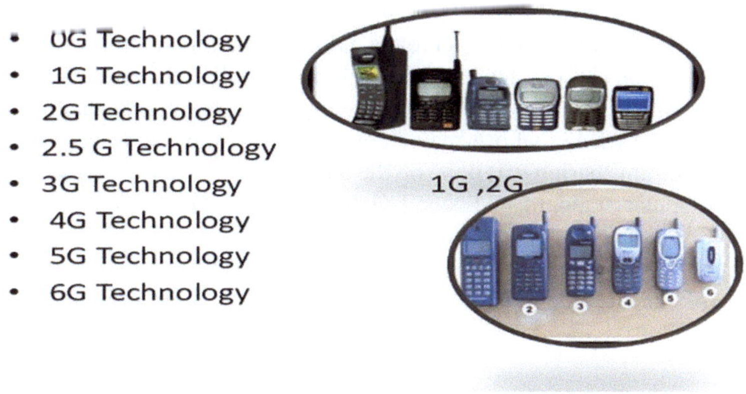

Fig. 1 Evolution of mobile communication

(MTS), Improved Mobile Telephone System (IMTS), and Advanced Mobile Telephone System (AMTS) which gave full duplex support and higher voice quality.

4.2 First Generation

Although 0G was a great change, it cannot be used by everyone, and another big step forward was needed, and it happened in 1979 when first-generation technology was introduced. First Generation of wireless mobile communication opened wireless doors to everyone and is meant for voice transmission only and is completely dependent on analog modulation technique. 1G was initially released in Tokyo, Japan, by Nippon Telegraph and Telephone Corporation (NTT), and within 5 years, Japan became the first fully wirelessly connected nation. 1G technology used analog signals for sharing information. The maximum speed in 1G was 2.4 Kbps. Spectrum sharing is done on the basis of FDMA here. The First Generation Technologies has certain disadvantage because it transmits analog signals, security measures are not provided. Any user who tunes to the frequency of the other user can hear the whole conversation. Thus, security measures are not provided here. Another disadvantage of this generation is spectrum inefficiency: if a user does not transmit any signal, then the spectrum allotted to that user goes to waste. Some of the standards of First Generation are NTS (Nippon Telephone System), NTT (Nordic Telephone and Telegraph System), AMPS (Advanced Mobile Phone Services).

4.3 Second Generation

The successor of First Generation was Second Generation (2G) which opened the doors for wireless communication to current advancements used for voice transmission with higher data rates of up to 64Kbps. 2G technologies used various standards: GSM, GPRS, and EDGE. 2G made use of digital modulation technique. The Second Generation (2G) systems were developed in 1991. 2G wireless mobile services gave some benefits over 1G services by giving additional facility of short message services (SMS) whose major emphasis was only on voice transmission services. Spectrum sharing was done on the basis of TDMA and CDMA techniques. The most popular standard of Second Generation was GSM (Global System for Mobile Communications) which was given by the ETSI (European Telecommunication Standards Institute). GSM used TDMA for spectrum sharing. The CDMA standard in Second Generation is called 2G CDMA or IS-95 [12].

4.3.1 2.5G

The next standard other than GSM of 2G was GPRS which was termed 2.5G which was developed in 1993 by ETSI and used packet switched network. It was called 2.5G instead of 3G because the 2G system was kept as it is and only some add-ons were added. The most common TDMA standard of Second Generation was GPRS (General Packet Radio Service). GPRS, also called as GSM-IP (Global System for Mobile Internet Protocol), is a packet switched system which means that all data in GPRS are transmitted and received in the form of packets. GPRS was meant for voice calls as well as data like accessing the Internet. GPRS led the upgrade of existing infrastructure rather than replacement. The GSM mobile of Second Generation were upgraded by a software update to support the GPRS infrastructure. As GPRS was meant for both voice calls and data, the data rate was quite high in GPRS as compared to GSM. GPRS is an easy billing system, and the billing is done on the basis of amount of data transmitted.

4.4 Third Generation

The user demands now increased from voice calls and data to other multimedia applications like live music, interactive web sessions, and conferences for which in the year 1998, Third Generation (3G) technology was developed. Data transmission speed in 3G was improved from 144Kbps to 2Mbps. 3G technology had also enlarged bandwidth and transfer rate so that web-based application and audio and video files can be accommodated easily. The standards of 3G were WCDM, UMTS, and HSPA. The International Telecommunication Union (ITU) decided that a common frequency of 2000 MHz will be used worldwide; hence, IMT (International Mobile Telephone System)-2000 evolved. The vision of IMT-2000 was:

- Common spectrum worldwide (1.8 GHz to 2.2 GHz)
- Wide range of telecommunication services (voice, data, Internet, multimedia)
- Data rates up to 2 Mbps
- Global seamless roaming
- Enhanced security and performance

The TDMA standards of 2G and 2.5G got converted to WCDMA (Wideband CDMA) in 3G, and CDMA standard of 2G and 2.5G got converted to CDMA-2000.

4.5 Fourth Generation

Fourth Generation (4G) of mobile communications was developed to increase data transmission speed, quality, and security. It is also called LTE which came in 2009. The data rate in 4G is 1Gps for stationary users and 100 Mbps for high-mobility users. Its wave frequency spectrum is between 900 and 2000 MHz. 4G also included another standard called WiMAX (Worldwide Interoperability for Microwave Access). WiMAX was standardized by IEEE; this standard is known as IEEE 802.16 which was released by South Korea in the year 2006. LTE as discussed was the appealing technology which is a packet switched system based on Internet Protocol. It uses Orthogonal Frequency Division Multiple Access (OFDMA) and Multiple-Input Multiple-Output (MIMO) concepts. 4G supports 3D television, IP telephony (VoIP), videoconferencing, mobile web, HD mobile TV, online gaming, and cloud computing.

4.6 Fifth Generation

At the time of 4G, people realized that they had what they needed from mobile telecommunication networks soon with the new emerging technologies such as IoT (Internet of things), smart cities with massive sensors, driverless chords, a more advanced technology after 4G was required and 5G appeared as a worldwide standard for air interface. It used the Time Division Duplexing principle and allots 1 Uplink time slot in view per each three Downlink time slots. The biggest change was Beam Forming and Massive MIMO, in comparison to the earlier technologies. It uses a very wide frequency range and it is for the first time that 5G has introduced various standards for different use cases. 5G uses the beamforming technique to attain exceptional selectivity. In more traditional networks, antennas are used to transmit data in all directions. All devices in a cell receive transmitted data though that was destined for some specific device. Depending upon the packet headers and address information, every device collects its specific data. In the case of beamforming, antenna arrays are used for transmitting signal at a desired angle. At any time of search, on the basis of availability of background messages in between antennas and devices, antennas try to identify the location of each device and direct information by only using the beam toward that device. Therefore, data transmission is done at a high speed and signal quality is maintained.

5G also uses Massive MIMO, which includes antenna set to transmit and receive signals and therefore increases reliability and throughput. Nowadays, data transmission and telecommunication are used in diverse areas with different requirements. In a few cases, data transfer amount is serious; in some other cases, the key issue is latency; sometimes, communication is in between a huge number of devices, while somewhere, only few devices communicate, but there is a requirement of high data rate. 5G has grouped all types of issues and divided them into three major categories:

- eMBB (enhanced mobile broadband): High amount of data is transferred, for example, videoconferencing.
- mMTC (massive machine type communication): The number of devices that need Internet access is high, for example, remote operation.
- URLLC (ultra-reliable low latency communication): Here, latency and reliability play a crucial role, for example, smart sensors.

The frequency range used in 5G is 400 MHz–1GHz, 6GHz, 30GHz, or 60GHz.

4.7 Sixth Generation

An innovative technology which seems to be revolutionary in the era of mobile communication with an objective to provide high data rates, i.e., fast speed for Internet access through mobile and wireless devices higher than 11Gbps, is the Sixth Generation (6G) [4]. It was beyond one's imagination to provide users Internet access at a very high rate, even if the user is moving in a remote location without any fluctuations. For this purpose, nano-antennas, which are specially developed, need to be installed at various topographical positions to be able to broadcast high-speed electromagnetic signals. Locations where nano-antennas can be placed are along roads, airports, hospitals, malls, villages, etc. Huge numbers of flying sensors will also be deployed all around with the support of 6G technology. These flying sensors will send information to their remote observer stations, and these stations will monitor activities in a special area like terrorist activities, burglary, etc. High-speed optical fiber lines assist point-to-point wireless communication networks which tend to transmit superfast broadband signals through the channel. Broadcasting of secured information from transmitters to destinations is done by airFiber. The radio over fiber structure is in running situation but with the introduction of 6G technology the earth's mankind will be very much nearer to any extraterrestrial refinement in our world. Major global companies are working on 6G now. The major applications of 6G include smart homes, robotics, cities, villages, space technology, self-driving cars, defense applications, home-based ATM systems, etc. Thus, 6G is the convergence of multiple technologies.

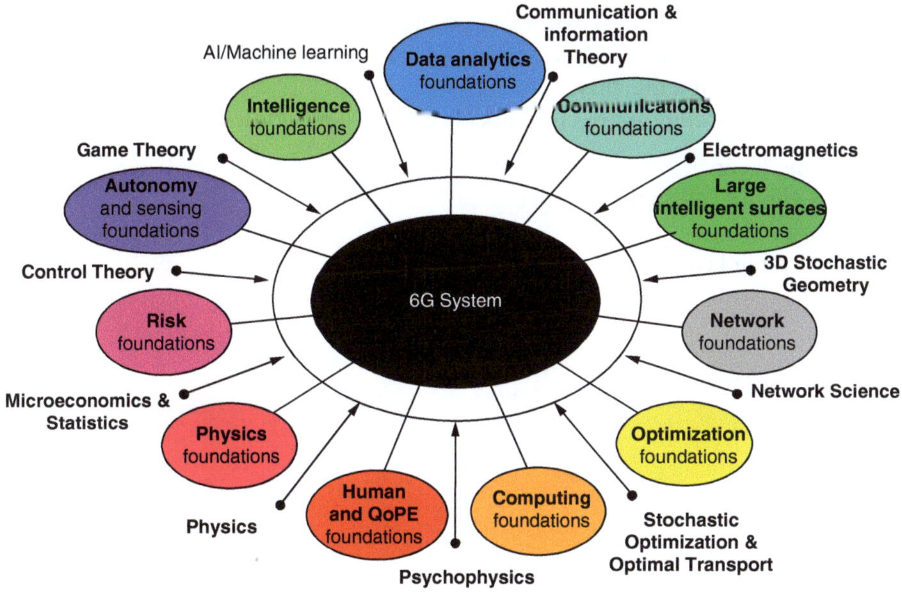

It will be further required by 6G networks to update themselves by understanding intellect to attain more precise desires and demands for the intelligent information society of 2030. The 6G technology will be integrated with ubiquitous AI with the intelligence of human mind inculcated in every aspect of networking systems.

5 Convergence of Ubiquitous Artificial Intelligence, Federated Learning, and 6G

The main objective of 6G technology is to achieve extremely high data rates up to 1Tbps and user experienced data rate of 1Gbps, very low latency with less than 1 ms end-to-end delay, ultrahigh reliability, higher energy efficiency, very high mobility, massive connection and traffic capacity, and large frequency bands. Thus, 6G networks will be required to transform themselves by recognizing intelligence.

6G networks to be developed will be highly complex, multi-layered, large scale, dynamic, and heterogeneous. These networks also need to support unbroken connectivity and also need to promise QoS desires of a large number of devices and also process a huge amount of data which is generated from physical surroundings. The requirements of 6G are:

- High-performance networking
- Higher energy efficiency
- High security and privacy
- High intelligence
- Increased device density

AI technology [1] with prevailing learning ability, ability for analysis, ability to optimize, and ability to recognition intelligently can be incorporated into 6G networks. These characteristics will make an impact to carry out performance optimization, discovery of knowledge, refined learning, group structuring, and complex decision-making intelligently.

The AI architecture for 6G networks is divided into four layers (Fig. 2):

- Sensing layer
- Data mining and analytics layer
- Control layer
- Application layer

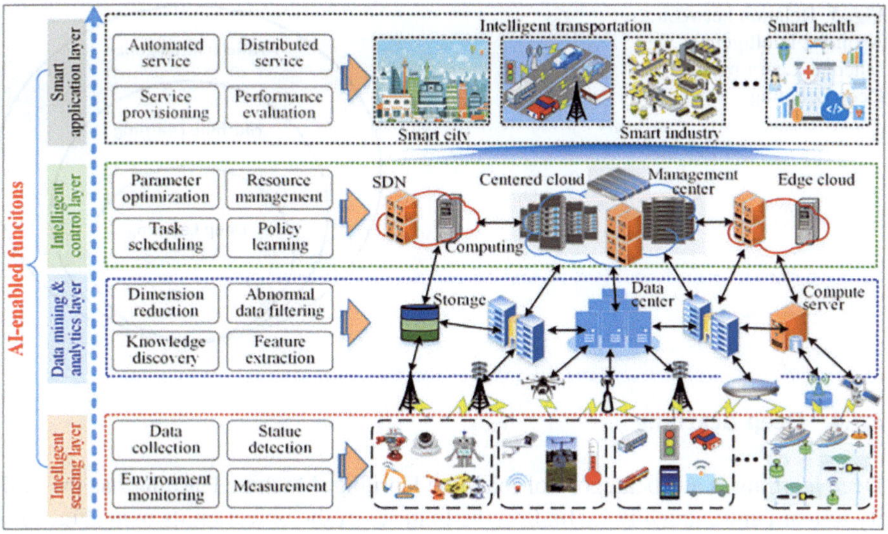

Fig. 2 Architecture of AI enabled 6G network

5.1 The Most Common AI Techniques Are Machine Learning (Supervised Learning, Unsupervised Learning and Reinforcement Learning) and Deep Learning

5.1.1 Supervised Learning

It is used to classify the dataset by learning the mapping function from the labeled dataset. It is broadly classified into:

- Regression: when the output variable is real number
- Classification: when the output variable is a category

5.1.2 Unsupervised Learning

This machine learning technique is used to identify similar groups of data, coined as clustering. The segregation of data is performed on unlabeled dataset, based on the inner structure of the data without looking into the specific outcome.

5.1.3 Reinforcement Learning

It is a machine learning algorithm which enables machines to improve its performance by automatically learning the ideal behaviors for a specific environment.

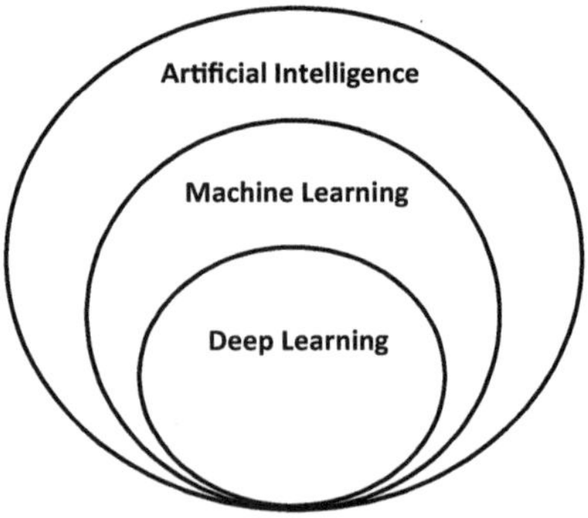

Fig. 3 Relation between Artificial Intelligence, Machine Learning and Deep Learning

5.1.4 Deep Learning

Deep learning is a subcategory of machine learning. Machine learning has the ability to learn automatically by finding the features of the object on its own (Fig. 3).

Now the conventional machine learning is centralized in data centers (clouds). The need of safety and the accessibility of huge data and computation resources in wireless networks are promising the developments of learning algorithms. This has led to the development of a fast-growing region known as federated learning (FL). Federated learning integrates two decoupled formerly fields: (1) wireless communication and (2) machine learning.

For all wireless devices, it is very impractical due to limited communication resources that are occupied in learning how to transmit whole gathered data to a data center that uses a centralized learning algorithm for data analytic or network self-organization. Therefore, it becomes increasingly attractive to process data locally at edge devices. This has led to the emergence of distributed optimization methods. In distributed optimization, each node can compute on its own data and send the results to its neighbors or a central node. Distributed optimization has many applications, such as user selection optimization, resource allocation optimization, trajectory optimization, and distributed machine learning design. Federated learning (FL) algorithm is one of the most emerging distributed learning frameworks which is also anticipated in future Internet of things systems.

In federated learning, wireless devices can supportively accomplish a learning task by only uploading local learning models to the base station in place of sharing the whole training data, as illustrated in the below figure. Since the data center cannot access the local datasets at the users, FL can protect data privacy of the users (Fig. 4).

For wireless communications, FL has the following advantages:

- Exchanging local machine learning model parameters instead of the massive training data saves energy and consumes less wireless resources.
- Training machine learning model parameters locally can effectively reduce transmission latency.

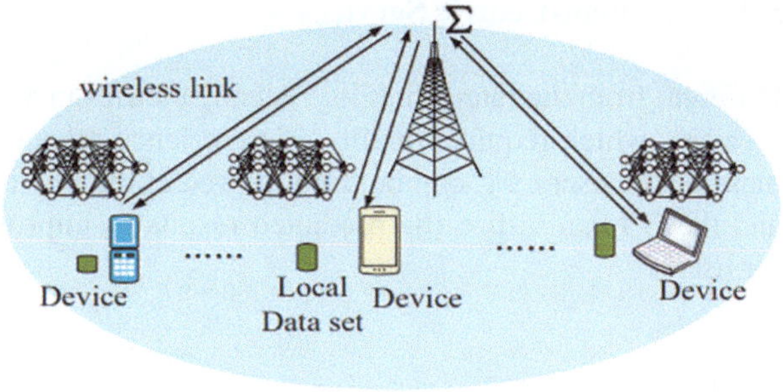

Fig. 4 FL algorithm over wireless communication systems

- FL preserves data privacy since the training data remains at each device and only the local machine learning model parameters are uploaded.
- Different learning processes are used which increases the likelihood of attaining higher precision.
- FL is inherently scalable since the growing amount of data may be counterbalanced by increasing the number of processors and providing a regular solution for large-scale learning where difficulty and memory are the major hindrances.

5.2 *Potential of Federated Learning to Support 6G*

It is anticipated that 6G communication systems can accommodate about 125 billion wireless devices by 2030. As a result, it is important to develop an automatic data processing framework to allow edge learning to take place. As one of the key enabling technologies, FL has the potential to meet the following 6G requirements [5].

5.2.1 Massive Ultra-Reliable, Low-Latency Communications (mURLLC)

Because of the explosive progression in the number of wireless devices in 6G, the 5G URLLC requirements will be changed to the mURLLC. With FL, multiple edge computing units can be used to cooperatively learn a shared model for the network, which can decrease service delay and provide high reliability.

5.2.2 Scalable Architecture

Different from a central cloud, edge intelligence, such as FL, is built in a distributed manner, which includes many edge servers with computing and communication capabilities. To serve a massive number of devices in the future 6G, it is essential to deliver a scalable and decomposable architecture to allow simultaneous computing among multiple edge servers. It is anticipated that the FL architecture will reflect an important and integral part in the forthcoming 6G services and applications [6].

5.2.3 Human-Centric Services

Different from the rate-reliability-latency metrics in 5G, 6G involves human-centric services, which requires quality of experience related to the physical eight movement of the users. FL can be used to predict the movements and gestures of users, and the BS can utilize the predicted results to improve the quality of experience for users.

6 Challenges and Future Scope [2, 3, 9]

A. *Practical Implementation of Federated Learning*

- *Expensive Communication:* In federated learning, more than thousands of devices are involved for model training; therefore, use of FL in 6G is a serious bottleneck. Several efforts have been done to improve communication efficiency, but for federated learning network, it is very challenging to coordinate it with local device. For efficient federated learning model which can be effectively used for 6G, substantial and diverse devices and networks are a requirement. The objective is to reduce communication overhead by reducing no. of gradients in each communication round and entire no. of communication rounds [7, 8].
- *Diverse Connectivity:* Modern experimental outcomes show that it is not required for individual device to apprise server in every iteration of model training. Federated learning is capable of converging impartial solution when all devices are equally likely. Practically in networking systems mobile gets frequent disconnection because of which substandard model or training result may be obtained. Therefore, the requirement is for such mechanism which can detect and monitor the connectivity status of all mobiles during model training.
- *Resource Consumption Optimization* [10, 11]: Two factors are there which are related to performance of federated learning: first is how much available and reliable is network connectivity, and second is computation capability of server and devices. When various artificial intelligence algorithms are applied, then resource consumption varies significantly. Therefore, quantification of resource consumption in different environment is very challenging.

B. *Artificial Intelligence Algorithm with Performance Assurances*

- *Artificial Intelligence with Worst-Case Performance Assurances*: When worst-case circumstances arise, modern networking systems are capable of offering some sustained services. Usually these worst-case circumstances are predicted; therefore, its impact on networking system can be evaluated in advance. It is foreseen that 6G is a relatively complex networking system; therefore, it is very difficult to estimate and evaluate possible consequences.

Developing a solution for evaluating network performance in different circumstances which offer worst-case performance assurance by using artificial algorithm is still an open and exciting problem.

- *Causal Inference in Complex Networks:* A number of artificial intelligence-enabled components may exist in case of 6G system which can work together. The major challenge is to preserve and correct such complex system to ensure that network has the capability to recognize input parameters which results in a precise output. Developing a casual interference-enabled artificial intelligence for 6G which is relatively complex is yet a very challenging job.
- Effective Issues: Deployment of federated learning models requires efficient training and inference. In case of low speed of model training, end users may not experience real-time services. When there is a requirement of deploying federated learning in 6G, some kind of challenges may be:
  - For a single device, size of federated learning model may be very large.
  - Model training may be very slow, so it is not possible to chase the delay necessities of 6G network.
  - Model inference may also be very slow because of which users' real-time request cannot be fulfilled.

7 Conclusion

This chapter has provided a comprehensive outline of communication networks from initial till 6G. Overview of ubiquitous artificial intelligence federated learning and fundamental requirements, challenges, and road map for 6G have been discussed. Latest requirements for 6G developments have been identified for which enabling technological solutions were discussed. Federated learning is a potential approach of machine learning which is capable of supporting the analysis of scattered data, and for mobile devices, their training at huge scale are proposed. Benefits, categories of federated learning, and performance indicators were presented. We discussed the motivating factors of federated learning along with its architecture which can be integrated with 6G to achieve secure, reliable, and efficient communication. Authors have tried a humble attempt of providing information so that it will spark additional interest for researchers to open a new research environment with the evolution of federated learning toward ubiquitous artificial intelligence and 6G.

References

1. P. Stone et al., "Artificial intelligence and life in 2030," One hundred year study on artificial intelligence: Report of the 2015–2016 study panel (Sep. 2016)
2. I. Stoica et al., "A Berkeley view of systems challenges for AI," Technical Report No. UCB/EECS-2017-159, University of California at Berkeley (Oct. 2017)

3. ITU-T, "Towards a new Internet for the year 2030 and beyond," ITU-T SG13 White paper (Jun. 2019)
4. S. Dang, O. Amin, B. Shihada, M.-S. Alouini, What should 6G be? Nature Electronics **3**, 20–29 (2020)
5. S.P. Yadav, K.K. Agrawal, B.S. Bhati et al. Blockchain-based cryptocurrency regulation: An overview. Comput. Econ. (2020). https://doi.org/10.1007/s10614-020-10050-0
6. F. Tariq et al., "A speculative study on 6G," arXiv preprint arXiv:1902.06700 (2019)
7. W. Saad, M. Bennis, M. Chen, A vision of 6G wireless systems: Applications, trends, technologies, and open research problems, IEEE Netw., 1–9 (20190
8. B. McMahan et al., Communication-efficient learning of deep networks from decentralized data, in Proceedings of the International Conference on Artificial Intelligence and Statistics, Fort Lauderdale, FL, 1273–1282 (Apr. 2017)
9. T. Li et al., Federated learning: Challenges, methods, and future directions, arXiv preprint arXiv:1908.07873 (Aug. 2019)
10. Y. Xiao, M. Krunz, Distributed optimization for energy-efficient fog computing in the tactile Internet. IEEE J. Sel. Area Commun. **36**(11), 2390–2400 (2018)
11. S.P. Yadav, Emotion recognition model based on facial expressions. Multimed. Tools Appl. (2021). https://doi.org/10.1007/s11042-021-10962-5
12. Y. Xiao, M. Krunz, Dynamic network slicing for scalable fog computing systems with energy harvesting. IEEE J. Sel. Area Commun. **36**(12), 2640–2654 (2018)

Communication-Efficient Federated Learning

Kaushal Kishor

1 Introduction

For decades, the exponential advancement of wireless networking methods along with various technical advances has vastly improved people's lives while also fostering business growth. The newest Fourth Generation (4G) LTE network, as seen in Fig. 1, has ushered in a modern period of mobile Internet, including core features including Web search, multimedia, and APPs [1]. The revolutionary 5G platform is intended to enable a greater range of providers, including augmented reality/virtual reality (AR/VR), the vast IoT (Internet of things), and self-driving cars [2]. The 5G infrastructure has three distinct technological features: improved cell wide band, machine-type huge coverage, and incredibly decreased communications. Businesses and academics have started to focus on the next phase of wireless infrastructure, coined as 6G [1, 2], 5G offers unrivalled opportunities to mankind and is now being implemented across the globe.

1.1 Why Is Federated Learning Important?

Traditional machine learning methods rely on centralized networks and are vulnerable to significant security vulnerabilities, such as a single point of failure. Furthermore, the researchers warn that consolidated data aggregation and distribution cause substantial overhead and that conventional centralized machine learning schemes may not be sufficient for 6G. This group includes federated learning (FL),

K. Kishor (✉)
IT Department, ABES Institute of Technology, Ghaziabad, UP, India
e-mail: kaushal.kishor@abesit.in

S. P. Yadav et al. (eds.), *Federated Learning for IoT Applications*, EAI/Springer Innovations in Communication and Computing,
https://doi.org/10.1007/978-3-030-85559-8_9

which is well-known for its autonomous machine learning approach. Two aspects contribute to the success of federated learning: its dispersed nature and the privacy it offers. Since AI, according to the developers, would connect human-centric growth with all aspects of network systems in a 6G world. As a consequence, 6G communications provide significantly higher levels of security and privacy.

1.1.1 Putting Federated Learning into Reality

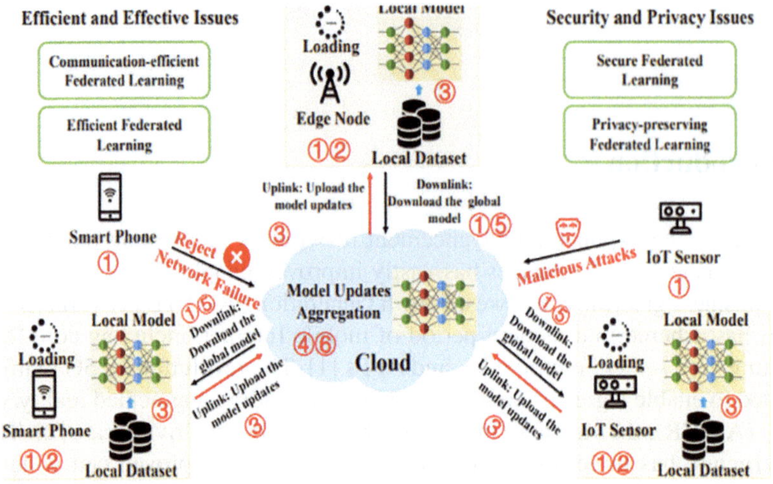

Fig. 1 Federal learning program implementation

1.2 Three Phases for FL-Based Architecture

1.2.1 A device's Service Requests May be Evaluated in the First Step of Initialization to See whether it Can Link with the Nearby Cloud to Train an ML Model through 6G

1. Each device can also receive initialized or pre-trained global models from the cloud.
2. In each iteration, the training process involves choosing each system and obtains the updated global model and trains a global model using a local dataset.
3. The cloud gathers model revisions from both devices that have been chosen and aggregates them to establish a new knowledge base over the next implementation, which is known as the aggregation method.

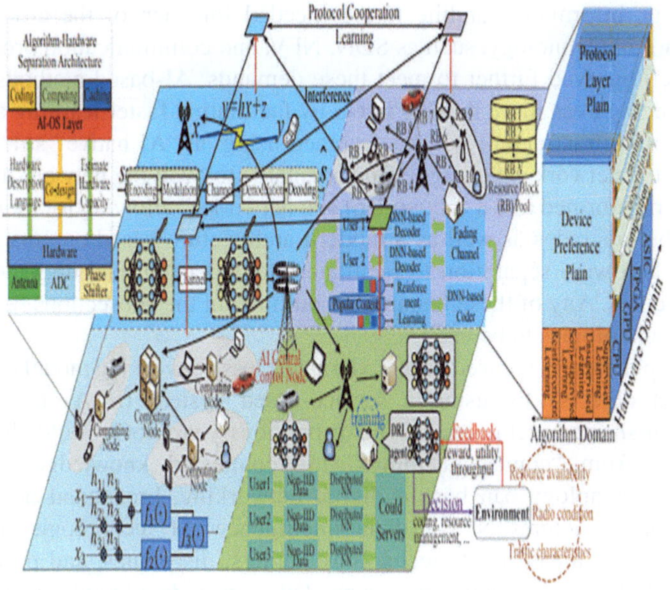

Fig. 2 The 6G architecture

1.3 The 6G Network Architecture

In this segment, we'll look at a possible 6G architecture (see Fig. 2) that incorporates network intelligence, sub-network evolution, and intelligent radio.

1.4 From Softwarization to Ntelligentization of Networks

We assume that 6G would bring network softwarization, like network knowledge, to new heights. The "non radio" aspect of 5G has grown in importance, and it has been the driving force behind recent "softwarization" efforts. Software-defined networking (SDN) and fully virtualized application are two main 5G technologies that have shifted modern telecom networks through software-based virtual networks. They also endorse wireless networks, an effective virtualization capability that allows various virtual networks to be deployed on top of a centralized physical infrastructure. Softwarization, on the other side, does not suffice for 6G as the network becomes more dynamic and heterogeneous. In order to assist AI-based networks, network providers must be able to execute a wide range of activities, including networking, knowledge caching, computing, and even wireless power transmission. 6G will allow modern radio networking frameworks such as THz connections and knowledgeable surfaces. Sensing, data processing, analytics, and storage are all advanced IoT features that will be needed. A scalable, agile, and, most significantly, informative architecture is needed for each of the aforementioned issues. Current technology, such as SDN, NFV, and communication protocols, will need to be improved further to meet these demands. AI-based methods can allow network slicing even more flexible in 6G by facilitating faster learning and adaptation. The 6G infrastructure will be designed using an "AI native" strategy, which will make the network smart, agile, and adaptable to evolving network dynamics. It would be transformed into a "network of sub networks," allowing for more powerful and scalable upgrades as well as a modern architecture based on smart radio and algorithm-hardware separation to facilitate interdependent and upgradeable hardware capacities. Any of these features, as outlined in the subsections below, would allow Ai technologies to be used.

The 5G platform is focused on a modern wireless networking model that achieves "connected objects" by using a service-based system (SBA) rather than an information-sharing architecture. Unlike previous years, 6G would shift wireless networking from "connected things" to "connected knowledge" [1]. Sixth-generation technology can benefit digital networking, broadband networks, and emerging infrastructures. New artificial intelligence (AI) technologies are likely to be used in these areas, as 6G technology gathers, distributes, and processes data quickly and effectively anywhere it is needed to provide a broad range of creative software and intelligent services [3]. Pervasive AI, in particular, would power 6G, a hyper-flexible architecture that builds all facets of networks using human-centric construction principles rather than data-centric, machine-centric, and application-centric concepts [4]. As a result, 6G interactions' security and privacy rights have been more strict. Traditional ML-enabled systems, on the other hand, that depend on a central server have major security issues, such as single turns of disappointment, and therefore cannot provide pervasive and stable AI for 6G. Furthermore, conventional clustered ML architectures might not be appropriate for ubiquitous ML due to the high overhead incurred by centralized data aggregation and delivery [5]. As a result, decentralized machine learning frameworks, which process all private data locally on training devices, are becoming more important for 6G. FL, a distributed machine learning address that is still in its infancy, has recently gotten a lot of support from academia and industry [2, 6]. FL participants use their local data to train a shared algorithm and make model changes to centralized parameter repositories rather than raw data [7].

FL is still in its development stage and is facing significant challenges in the 6G space, such as its excellent vision for AI-powered 6G and significant advancements in privacy-sensitive 6G applications. The Core federated learning (FL) challenges in 6G communications, such as high transmission costs due to numerous contact stages for method adjustments and accumulation, are first addressed [8]; poisoning risks and backdoor attacks are examples of security problems generated by heterogeneous and disparate participating entities [7, 9] and privacy concerns triggered by gradient leakage attacks [10]; and mega-scale massive-scale massive-scale massive-scale massive-scale massive-scale massive-scale massive-scale massive-scale massive-scale massive-scale massive. Then, to address the above issues from a variety of perspectives, we propose novel federated learning approaches. Finally, we talk about FL's future 6G communications paths as well as versatile research topics.

1.5 6G for Applications in Artificial Intelligence

A slew of AI-enabled smartphone applications have arisen as a consequence of the rise of the emergence of AI and knowledgeable mobile devices. In this part, we'll show you how 6G can affect mobile AI apps. Since AI tasks involve a lot of computation, they are normally prepared, installed, and installed in data centers on custom-built servers. With the widespread proliferation of smart mobile devices, a huge proportion of application domains are expected to be launched at the Wi-Fi network's edge. As a consequence, using advanced wireless networking and mobile computing technology, the 6G wireless network would be built to support various network mobile devices with AI-enabled implementations. Wireless network intensity and data rate are the primary slowdowns in mobile application domains for three reasons: First, to protect privacy, certain AI programs need data to be saved on mobile devices rather than sent to the cloud for model testing [11]. Latest research has piqued curiosity in on-device distributed instructions. Second, by pooling the computation and storage ability of several cell units and addressing the resource constraints of edge computers, on-device distributed computing opens up new possibilities. In this case, data shuffling is a critical part of exchanging measured intermediate values between mobile devices [12]. Finally, a distributed processing device for DNN testing and inference is created using a heterogeneous combination of cloud, peripheral, and end machines.

6G will provide scalable architectures for implementing new networking and computing technology, allowing for pervasive and diverse broadband AI networks. Machine learning frameworks that are distributed communicate with each other. We'll look at how 6G can solve convergence problems for large-scale distributed machine learning in mobile AI applications in this segment. On-device distributed training can now be achieved by processing data locally, thanks to the increasing computing and storage capacity of computers. For dispersed training on mobile devices, contact via the volatile wireless medium, on the other side, becomes a serious bottleneck. Federated learning [11] enables remote mobile devices to study a single global model by processing training data globally on each computer, improving data privacy and protection. The minimum for certain applications, bandwidth has become a barrier. Model accumulation estimated from regionally modified models tabulated at each separate terminal, however. It is necessary to do simulations over its air used to allow relatively low model aggregation by exploiting the entanglement property of an ethernet standard of transparency, as shown in Fig. 3. Joint system selection (i.e., maximizing the number of chosen devices) and beam forming configuration (i.e., minimizing the model parameter aggregation error) are being used to improve the transient response, the consistency rating in the estimation step, and the decentralized teaching strategies.

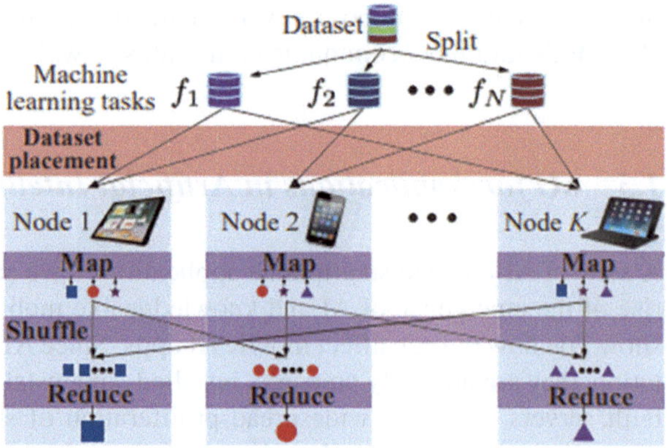

Fig. 3 Wireless map reduce on-device distributed inference

1.5.1 Intelligence around Applications

- At the moment, 5G network implementations are becoming rapidly intelligent. Intelligent systems are one of the application standards for 6G networks [9, 13]. FL allowed wireless networking technology to link smart phones to 6G networks, allowing them to operate a multitude of intelligent applications. Users may require intelligent Google Assistant in the future to accomplish their daily routines, for example [14]. Users can benefit from highly intelligent software, thanks to the 6G network's pervasive AI.
- Intelligence on service. In addition, as a human-centered network, the 6G network's high awareness will offer appropriate and personalized intelligent services [1, 2, 4]. For example, Florida utilizes distributed learning to deliver customized healthcare, personalized recommendations, and personalized intelligent voice services to its customers. In the future, intelligent devices would be tightly linked to 6G networks [4].

1.6 Deep Sea and Space Communication

In terms of both economic and technological opportunities, space tourism has enormous potential in the coming decade [15]. Humans from all walks of life can venture into space. For space exploration, some companies are preparing to fly sub-orbital commercial flights. Following a few efficient and lucrative space flights, the next move would be to ensure that consumers have access to space hotels and space hospitals [16]. Aside from commercial space travel, another possible use is space exploration [15].

Through the availability of simple and reliable networking methods, 6G can broaden the number of activities available across the world. For coordination and scientific purposes, autonomous and knowledgeable robotics can be used in harsh environments [16]. With the strong and massive capacities of 6G networks, the world's puzzles could be solved. Deep-sea drilling for oil and minerals, for example, may become a possibility.

1.7 Networks that Are Energy Consuming

Huge self-organizing and self-healing robotics can make up the next-generation wireless networking infrastructure. Both of these intelligent robots/devices would necessitate a lot of computing capacity. As a result, as the number of intelligent robots grows, so will the need for electricity. Traditional GPUs are currently incapable of achieving the energy consumption standards of next-generation wireless networks. Energy-efficient and scalable intelligent network architecture would be expected in such a scenario. IoTs, IoEVs, and IoBTs have been common in the industry [17, 18]. We have sensors all over the place. A sensor can be found in our entrance, air conditioner, vehicle, television, refrigerator, and workplace. Many of these sensors need low-energy contact [19, 20].

2 Overview and Preliminaries

Unlike 5G, 6G has influential features that guarantee universal, high-performance convergence, smooth, and intelligent networking while maintaining security and privacy. We'll go over the efficiency specifications for 6G communications in more detail below.

2.1 High-Capacity Networking

Many people believe that 6G is a complex networking infrastructure comprised of multiple disparate earth orbit transmission networks [2, 4]. Three-dimensional amazingly networks provide share in the world and enhanced networking to encourage various forms of security solutions and dense availability via subsystems and subnetworks including such underwater-land networks and satellite networking devices. 6G communication channels will actually achieve a data rate of up to 1 Tbps per user, with incredibly low edge latency, excellent end-to-end effectiveness, and high power consumption networking with the help of tremendous load balancing [2]. 6G communications, in contrast to 5G communications, include sustainable and low networking and compatibility not only in dense urban areas but also in less developed regions such as underwater environments [4]. In the face of virtual networking and interaction from everywhere, 6G communications allow use of cutting-edge connectivity networks to include a broad variety of data, including audio, video, and augmented reality (AR/VR), culminating in a new touch interface [2].

2.2 Energy Efficiency Improvements

In the age of 6G, communication systems will have relatively high energy conservation ratios, charging limits, and power consumption restrictions. As a result, two common 6G communication systems study had provided battery life is extended and energy demand is reduced. To solve the power crisis of wireless devices, particularly smart phones, existing works have suggested energy storage devices. To improve energy efficiency life and operating time of connected networks, especially smart phones, researchers are using energy-harvesting technology [4]. Wireless meters, in specific, will collect electric power from radio-frequency, solar, geothermal, and wind energy and store it in a variety of energy storage systems, extending battery life [21]. Wireless networks with inductive charging functionality, such as unmanned aircraft (UAVs) and electric vehicles (EVs), can also receive energy supplements via wireless communication technologies from complicated network infrastructures or mobile charging stations [22].

AI-based technologies are critical for green networking approaches to optimize energy consumption and wireless energy management networks in a complex setting of several performance targets. Deep reinforcement learning and other sophisticated machine learning methods may be used to improve a wireless device's computational task offloading decision and find the best working and sleeping time scheduling solution, resulting in lower energy consumption and improved efficiency. In 6G network densification scenarios, AI-based deployments would benefit multi-hop communication routing, collaboration relay availability, and networking strong preference, as they greatly reduce wireless network data transmission while significantly improving communication [2, 4].

2.3 High Levels of Safety and Privacy

The majority of recent 4G and 5G research focuses on network throughput, reliability, and latency [4]. Reliability and privacy issues in wireless networking, on the other hand, have received little attention in recent decades [23]. Protecting data protection and privacy has been a core aspect in human-centric 6G communications because data protection and privacy considerations are very tightly linked to users' lives. Meanwhile, communication/data service providers are required by law to collect a large amount of user data, resulting in frequent privacy data leakage. To solve this problem, in 6G networks, FL methods must be used to achieve secrecy deep learning.

2.4 *Superior Intelligence*

Users can benefit from 6G's high intelligence by receiving high-quality, customized, and insightful services. Operational intelligence, program intelligence, and information program are also used in high-intelligent 6G.

2.4.1 Operational Intelligence

Problems including resource sharing and multi-objective output management are common in traditional network operations [1, 24]. To achieve a suitable level of network activity, optimization approaches like game theory, contract law, and other theories are popularly used. However, in size of an organization numerous factors and multi-objective scenarios, these simulation theories can struggle to make the correct decision. The above can be solved using highly evolved machine learning approaches due to advances in deep learning methods. The emergence of federated learning, but from the other end, has turned multi-objective linear classification problems into nonlinear optimization problems, enabling the best way for complex and time-varying information management decisions [2].

2.5 *Scenarios in Typical Use*

The 5G service concept has evolved into an application-based infrastructure in compared to older years, with advanced mobile broadband, tremendous machine-to-machine communication, as well as ultra-reliable, low-latency communication as system examples. 6G will enable the newly added service categories, as shown in Fig. 2, thanks to Industry 5.0 and deep learning technology.

- **New Media –** With the fast expansion of cellular networks, networking technology, it is planned to happen community engagement will shift from AR/VR to extended truth in high fidelity (XR) touch during the next 10 years, paving the way for wireless holographic communication [1–4]. At any time and from any location, users can access existing holographic communication and show services. Digital education, virtual travel, virtual athletics, virtual sculpture, virtual concerts, and other fully immersive holographic events are only some of the possibilities [25].
- **New Services –** As per the ITU-6G T's networking technology journal article [1–4, 26], further and moderate connect technology can provide potential customers including a broad selection of modern services. Quantum networking, holographic teleportation, visible light connectivity, and several other communication new technologies have cast doubt on the model of conventional operation. Digital surgery [27], cloud PLC [28], and intelligent transport infrastructure [29]

are numerous companies that could benefit from the new software platform. The aim of this new technology is to create facilities that are elevated, mutable, and strong.

- **Evolving Infrastructures –** Thanks to deep learning advances, the 6G networking framework has spawned a slew of new infrastructures, including integrated terrestrial and space network [27], federated learning networks [11], centrally controlled infrastructure and services [1], and trustable infrastructure [4]. The FL network, in particular, benefits from the 6G network's high bandwidth and low latency, which has caused a slew of new smart devices to be deployed to cities, factories, and citizens [27].

3 Federated Learning

In this section, 6G will be referred to as a distributed learning system based on FL. A large number of distributed machines linked to different utilities will cooperate to train a common global model leveraging locally collected datasets (e.g., anomaly detection, suggestion framework, prediction of the next term, etc.). Figure 4 depicts the FL-based modelling method's initialization, making plans, and aggregation phases. A device may investigate customer demands, specifications, and network security criteria mostly during the initialization phase to determine whether it should register with the nearest cloud for joint global training stage over

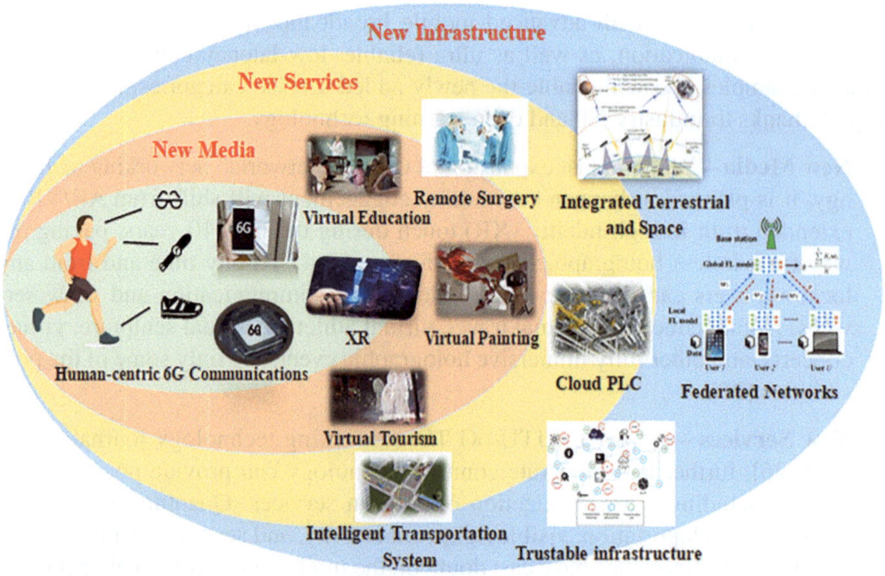

Fig. 4 Key 6G services and a roadmap [16]

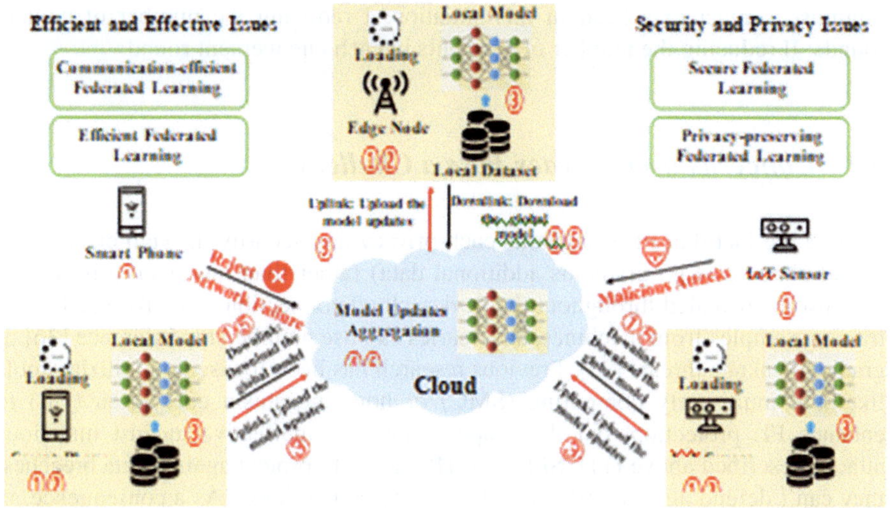

Fig. 5 Summaries of 6G federated learning mechanisms [17]

a wired or wireless link (e.g., 6G). In its role as task developer, the cloud would then select at random a subset of the registered devices to participate in the next round of training, ignoring the phones that are already recorded. Frequently, a cloud-based global model that has already been pre-trained or prefixed is sent (Fig. 5).

4 Primary Difficulties in Federated Learning for 6G

The key problems of FL, which are the key bottleneck issue prior to every large-scale implementation of 6G implementations, are discussed in this portion.

4.1 Expensive Communication Is a Challenge

Touch is important with 6G, and there is a bottleneck for FL [11], since FL requires thousands of devices to engage in model training. Previous research [6, 8, 10, 30–34] sought to increase the communication performance of the FL method in a number of ways. Furthermore, for FL networks [8, 10, 35], achieving connectivity that is compatible with the device's local measurement is difficult. To deploy the FL method to huge 6G networks, diverse devices and networks, an information exchange solution that can vastly decrease the consumption of patterns switched back and forth computing and cloud computing rather than always gradient information must be created. Two main factors must be weighed in order to further minimise contact over-head in this situation: I reducing the number of contact rounds; II reducing the number of gradients in each engagement round.

4.2 Concerns over Privacy Pose a Challenge

Despite the fact that FL safeguards each privacy and security through exchanging model changes (e.g., gradients additional data) rather than raw data, private data will also be revealed throughout the device-cloud connection [36]. To snatch local training samples from machines, adversaries can use membership inference [36] or gradient leakage threats [37]. Previous research has been focused on utilizing tools like safe multi-party computing (SMC) or homomorphism encryption (HE) to enhance FL protection, but these approaches are ineffective against malicious attacks described above [11]. SMC and HE can only protect against data breaches; they can't defend against inference attacks or gradient leaks. As a consequence, in order to resist or mitigate the above destructive assaults, the FL device must introduce new privacy-enhancing measures as quickly as possible.

4.3 Effective Issues Are a Challenge

Despite the fact that FL safeguards privacy and security through exchanging model changes (e.g., gradients details) instead of just datasets, private data will also be revealed throughout the device-cloud connection [36]. To snatch local training data from machines, adversaries can use membership inference [36] or gradient leakage threats [37]. Previous research has concentrated on using tools like stable multi-party computing (SMC) or homomorphism encryption (HE) to improve FL protection, but these approaches are inadequate against the malicious attacks described above [11]. Only SMC and HE can avoid identity theft; model preparation and inference are typically needed when deploying FL models to computers [11]. If there is model planning and thought are soft, consumers would not be able to gain from relatively clever networks [38]. As a result, if FL devices are widely deployed with 6G technology, the following concerns will arise: FL model calculation is too slow to meet the user's need; FL model testing is too slow to fulfill the intensity demands of the 6G network; and FL method inference is too weak to meet the customer's vast competition. For the seamless integration of FL and 6G networks, appropriate equipment and assumption are needed a diverse, heterogeneous context, FL frameworks, but in the other hand, struggle to achieve effective model training and inference. They can't ward off attacks like participant inference or gradient leaks. As a consequence, in order to resist or mitigate the above destructive assaults, the FL device must introduce new privacy-enhancing measures as quickly as possible.

5 For 6G, Advanced Federated Learning Methods Are Accessible

We suggest advanced federated learning systems that use a range of evolving innovations or methods that make for more effective contact, robust, and privacy-enhanced federated learning to address the aforementioned issues.

5.1 Federated Learning with High Communication Efficiency for 6G

It's difficult for devices that cover a greater geographic region to achieve a better global model in 6G, although it's possible with a lot of connectivity overhead. The connectivity overhead would have an impact on the gradient sharing between the machines and the server, which will have an impact on model aggregation in the cloud. As a result, we must devise a more effective method of achieving FL preparation. In this part, we'll go over information exchange FL from a computer and algorithm standpoint, paving the way for wider FL adoption and use in 6G communications.

5.1.1 Effective Communication FL

Data propagation (e.g., non-independent and equal dissemination), system delivery (e.g., heterogeneous machines across areas and networks), computation systems (e.g., dispersed and centralized), and synchronization techniques (e.g., synchronous and asynchronous scheme) all have different effects on communication performance of different implementations [8].

5.1.2 Effective Communication FL

By hastening convergence [35] and dropping the communication expense of each round with gradient compression methods [10], building Information sharing Protocols at the Algorithm Level FL would minimize the amount of modifying an existing necessary to train a model.

5.1.3 Model Convergence Acceleration

To minimize the number of training set iterations, stochastic gradient descent algorithms based on zero-order, first-order, second-order, and federated convergence are used [8]. The federated optimization approach is very famous for this specific opportunity in accelerating training model convergence since it can secure the private data on each system.

5.1.4 Reducing Connectivity Overhead

In order to reach communication-efficient FL, gradient scarification and quantization will significantly reduce the vast amount of gradients shared between the machines and the server. Lin et al. suggested a gradient focused on top-k range compression technique in [10] to increase the effectiveness of contact. The authors compressed the differential 300 times in this procedure to lessen the quantity of gradients while maintaining precision.

5.2 6G Safe Federated Learning

FL would be prone to malware attacks from heterogeneous networks owing to the large spectrum of 6G network links, interdependent machines, and malicious participants in the training process [9]. Researchers have suggested a range of security techniques to approach this topic from three angles: aggregation algorithms, detection systems, and credibility management.

5.2.1 Algorithm for Robust Aggregation

Aggregation is a key phase in the FL training method that affects model convergence significantly. The robust aggregation algorithm was developed to counteract the negative effects of low-quality model modifications triggered by means of malicious software (such as confounding threats) on worldwide model training. The cloud will be able to tolerate Byzantine device failures if this strategy were used [39, 40]. Ang et al. suggested the regularize approximation approach in [41] to minimize noise distortion in heterogeneous systems and networks.

5.2.2 Detection Mechanism of High Efficacy

Another obvious approach is to classify suspicious equipment and prohibit them from taking part in FL preparation. Traditionally, the accuracy of the computer's sub-model has been used as an assessment parameter for detecting malicious devices. To address the malicious attack issues, Liu et al. used blockchain smart contract techniques to create a malicious device detection Scheme [42].

5.2.3 Reliable Reputation Management

A measure called reputation may be used to assess the efficiency and trustworthiness of systems based on their past habits. The more credible the unit is, the more reliable it is. As a result, building a credibility protection system for system historical activities in FL may also protect the global model from malicious applications. Kang et al. suggested a credibility management system in [23], and local differential privacy (LDP) strategies accomplish this aim by manipulating device gradients [36]. DP and LDP technologies, on the other hand, improve FL privacy at the cost of model efficiency. As a result, innovative approaches that combine privacy and efficiency are currently available, as mentioned below.

5.2.4 Neural Network Pruning

A deep learning technique called neural network pruning aims to make a neural network simpler and more effective. Huang et al. [43] used pruning as an alternative to DP to preserve the FL system's privacy while maintaining model efficiency. This method incorporates a modern approach to view the model pruning to produce the same outcomes as DP techniques, opening up new avenues for balancing utility and privacy.

5.2.5 Compression of Gradients

The gradient information includes rich textual information, which allows adversaries to conclude the local data of the machines [37]. Methods that disrupt the propagation of gradient information protect gradient secrecy, according to an intuitive theory based on the above. Gradient compression will protect against gradient leakage attacks without losing precision, according to Zhu et al., and the security impact is better than DP.

5.3 For 6G, Effective Federated Learning Is Important

Human-centric connectivity services in 6G networks provide a long-term objective of handling customer needs in real time. As a result, effective FL must be achieved by preparation and inference.

5.3.1 Effective Training

Efficient training will drastically reduce the amount of time it takes to train mobile devices for productive FL. The following is a list of specialized teaching techniques.

Federated Parallelization

To speed up model training, data and model parallelization is often used. By running several training samples in parallel, data penalization achieves effective training [44, 45]. Model parallelization divides the model through several processors to speed up model training [38].

Federated Distillation

The performance of a pre-trained complicated system (the Instructor model) is used as a guided signal to train another accordance with the established in model distillation (the Student model). Using student models, this approach may be used to enhance model training efficiency. Jeong et al. proposed federated distillation (FD) in [32], an algorithm for training distributed models that outperforms the FL prototype scheme, notably when the prototype size is large.

5.3.2 Efficient Inference

The current FL model is much too broad to enable computers to make real-time inferences. To achieve effective inference, use the methods mentioned below.

Pruning

Pruning is a model optimization procedure in which weight tensor and excess weight are removed. The lightweight neural network not only runs quicker, but it also lowers the cost of computing for the proposed method, which is an important move in bringing the model to smart phones and other user equipment.

Weight Sharing

Weight sharing reduced the quantity of parameter values, allowing model prediction to be more precise. The explanation for this is that the bigger the model is, the less parameter it has. Weight sharing strategy for wireless networks used by Tran et al. in [33] to increase model inference reliability has to be solved immediately. In their pioneering work, Li et al. introduced q-Equal FL (q-FFL), a modern aggregation algorithm for achieving a balanced allocation of capital and precision [46].

5.4 Federated Learning May be Explained

Blackbox configurations make up the overwhelming majority of FL models (i.e., non-interpretable), making it difficult for users to grasp what tools the model has. In a dynamic 6G network infrastructure, the blackbox model's unexplainable predictions or actions could result in significant losses for subscribers. To assess when the car is moving or halted, 6G-enabled self-driving, for example, uses a visual perception model for in-vehicle use. Since the on-vehicle model is not interpretable, the driver is unable to comprehend the vehicle model's production judgment. Uber's self-driving car suffered a car crash in 2018 when it was programmed incorrectly.

6 Future Research Directions and Open Research Topics

6.1 Federated Learning that Is Reliable

6.1.1 Federated Learning with Privacy Enhancement

Previous FL research focused on consumer or cloud-level anonymity for all users in 6G networks. The previous schemes, on the other hand, place strict privacy limits at the expense of accuracy [11]. Since the market is concerned about the FL model's accuracy, it's critical for FL to introduce privacy-enhancing techniques that don't compromise accuracy in exchange for stringent privacy guarantees. At this end, just a few researches are looking at possible alternatives. Huang et al. [43], for example, recently introduced a net pruning technique that replaces pruning with the DP technique to include stringent privacy safeguards while still optimizing the training efficiency of the model. Designing methods that can balance durability and privacy restrictions in reality is an interesting and ongoing path.

6.1.2 Federated Learning with Improved Security

FL applications are susceptible to disruptive attacks by adversaries against various organizations because they often include several entities such as smart phones, cloud, and deep learning model providers. While there has been a lot of effort put into providing good security protection for FL applications, there has been little effort put into defending or mitigating these malicious attacks from a device standpoint. In [34], Bonawitz et al. investigated more efficient and stable aggregation algorithms and fault tolerance mechanisms from a machine design viewpoint. FL is developing more realistic industrial technologies with the aid of 6G networks, so the security-enhanced approaches are developed from a device viewpoint.

6.1.3 Fair Federated Education

FL entails thousands of devices in large, heterogeneous networks training a shared global model [46]. In such a network, a naive optimization of the global model may be unreasonable to certain systems, resulting in unequal advantages or disadvantages. Naturally, FL toward justice is a prerequisite for human-centric 6G connectivity services. In a wireless network, a fair FL implies fair resource sharing and a rational compensation mechanism. How to efficiently and equally distribute computational and communication capacity in large, heterogeneous networks has become a crucial task. In the sense of a complex network structure, such as 6G, the implementation of an interpretable FL model is required to provide human-centric communication services.

6.2 Federated Learning that Is both Efficient and Effective

- Transmission overhead is still a bottleneck in commonly used FL systems, notwithstanding the fact that the 6G network will provide FL systems with relatively low latency [11]. The method of sequential batching and the asynchronous batch approach are also accessible (where the model update latency is supposed to be bounded) [34] are the two most widely discussed in decentralized machine learning applications, connectivity modelling schemes are used as defined in Sect. IV-A1. Since asynchronous synchronization strategies have included a scheduler, supervisor, workers, and auto-update, there would be multiple optimization problems to solve in the future for some of these positions: how the scheduler uniformly distributes the communication and processing energy of the system, how the organizer efficiently governs the operating and idle states of the machines, how the worker efficiently tracks the working and idle levels of the computers, and how the worker efficiently regulates the working and inactive states of the devices. Potential studies can concentrate on these optimization problems in order to construct innovative asynchronous structures.
- Since most FL models have a pre-defined framework, this architecture is perhaps not the best choice for non-independent and equivalent delivery (non-IID) performance. As a result, automating FL (AutoFL) schemes based on neural architecture search (NAS) may be a potential solution to this issue. A federalized NAS (FedNAS) algorithm was introduced to aid distributed machines in organizing to find a better, more precise framework. In the future, NAS will make it possible to create a stronger FL model architecture [36].

6.3 *The Direction of Incentive Federated Learning*

The majority of current research focuses on improving FL algorithm performance, such as precision and training time. In 6G circumstances with massive self-interest computers, however, it is unrealistic to assume all data owners to be able to join the FL at any time and from any place. As a consequence, reward mechanisms for truthful and productive engagement have emerged as an important and pressing research subject [7, 18, 46–49]. Designing successful online-learning-dependent reward approaches to reduce the impacts of both knowledge asymmetry and time-varied usable tools is also an open issue associated with information asymmetry between competition publishers and competing computers, such as details regarding time-varying accessible services, unfixed operating hours, and constantly changing participation willingness. On the other side, data consistency has a huge influence on learning outcomes. Creating high-quality information reward strategies to enable more materials with enhanced data in order to take part in federated learning and gain higher remuneration for their strong data inputs, thus improving machine productivity and learning capacity, is a difficult problem to solve [7, 46, 48].

6.4 *Steps toward Customized Federated Learning*

It is difficult for the FL device in the 6G network to create customized services to consumers. Prior investigation [50–53] using a series of tailored methods to offer real-time personalized services to customers, and is a significant step toward personalised FL. However, non-IID info, machine heterogeneity, and network heterogeneity continue to pose challenges to customized FL. Personalization is a critical component of human-centric 6G networks. As a result, finding new approaches to overcome the above issues is a fascinating and worthwhile topic for FL to pursue.

7 Conclusion

In this part, we summarized how to incorporate federated learning with 6G communications. The core difficulties of federated learning for 6G applications, as well as the requirements of 6G networking, were discussed. We presented a quick review of the evolving innovative methods to federated learning for 6G communications, comprising information exchange federated learning, stable federated learning, and effective federated learning, to overcome the above challenges. Finally, we established a range of open topics and research paths that should be pursued further.

References

1. K.B. Letaief, W. Chen, Y. Shi, J. Zhang, Y.-J.A. Zhang, The roadmap to 6g: Ai empowered wireless networks. IEEE Commun. Mag. **57**(8), 84–90 (2019)
2. Y. Xiao, G. Shi, M. Krunz, Towards ubiquitous AI in 6g with federated learning. arXiv preprint arXiv, 2004.13563 (2020)
3. K. David, H. Berndt, 6g vision and requirements: Is there any need for beyond 5g? IEEE Veh. Technol. Mag. **13**(3), 72–80 (2018)
4. S. Dang, O. Amin, B. Shihada, M.S. Alouini, What should 6g be? Nat. Electron. **3**(1), 20–29 (2020)
5. S. Niknam, H.S. Dhillon, J.H. Reed, Federated learning for wireless communications: Motivation, opportunities and challenges. arXiv preprint arXiv, 1908.06847 (2019)
6. J. Konecˇny`, H.B. McMahan, F.X. Yu, P. Richta'rik, A.T. Suresh, D. Bacon, Federated learning: Strategies for improving communication efficiency, arXiv preprint arXiv:1610.05492 (2016)
7. S.P. Yadav, K.K. Agrawal, B.S. Bhati, et al., Blockchain-based cryptocurrency regulation: an overview. Comput. Econ. (2020). https://doi.org/10.1007/s10614-020-10050-0
8. Y. Shi, K. Yang, T. Jiang, J. Zhang, K.B. Letaief, Communication-efficient edge ai: Algorithms and systems. arXiv preprint arXiv, 2002.09668 (2020)
9. Y. Liu, J. Peng, J. Kang, A.M. Iliyasu, D. Niyato, A.A.A. El- Latif, A secure federated learning framework for 5g networks. arXiv preprint arXiv, 2005.05752 (2020)
10. Y. Lin, S. Han, H. Mao, Y. Wang, B. Dally, Deep gradient compression: reducing the communication bandwidth for distributed training, in International Conference on Learning Representations, 2018. [Online]. Available: https://openreview.net/forum?id=SkhQHMW0W
11. A. Jain, K. Kishor, Financial supervision and management system using ml algorithm. Solid State Technol. **63**(6), 18974–18982 (2020)
12. B. McMahan et al., Communication-efficient learning of deep networks from decentralized data. Proc. Int'l. Conf. Artificial Intell. Stat. (AISTATS) **54**, 1273–1282 (2017)
13. T. Li, A.K. Sahu, A. Talwalkar, V. Smith, Federated learning: challenges, methods, and future directions. IEEE Signal Process. Mag. **37**(3), 50–60 (2020)
14. J. Kang, Z. Xiong, D. Niyato, Y. Zou, Y. Zhang, M. Guizani, Reliable federated learning for mobile networks. IEEE Wireless Communicat. **27**(2), 72–80 (2020)
15. R. Moro-Aguilar, The new commercial suborbital vehicles: An opportunity for scientific and microgravity research. Microgravity Sci. Technol. **26**(4), 219–227 (2014)
16. S.S. Hassan, C.S. Hong, Network utility maximization for 6G maritime communication in deep waters. J. Korean Informat. Sci. Soc., 957–959 (2019)
17. B. Sliwa, R. Falkenberg, C. Wietfeld, Towards cooperative data rate prediction for future mobile and vehicular 6G networks, in *2020 2nd 6G Wireless Summit (6G SUMMIT)*, (IEEE, 2020), pp. 1–5
18. Y. Qian, M. Chen, J. Chen, M.S. Hossain, A. Alamri, Secure enforcement in cognitive internet of vehicles. IEEE Internet Things J. **5**(2), 1242–1250 (2018)
19. J. Scott, A. Stevenson, H. Lupa, Space tourism: An acceleration physiologist's perspective. Aviat. Space Environ. Med. **83**(3) (2012)
20. N. Henbest, Private space travel: diary of an astronaut in waiting. New Scient. **220**(2944), 41–43 (2013)
21. K. Kishor, P. Nand, P. Agarwal, Subnet based ad hoc network algorithm reducing energy consumption in manet. Int. J. Appl. Eng. Res. **12**(22), 11796–11802 (2017)
22. K. Kishor, P. Nand, P. Agarwal, Notice of retraction design adaptive subnetting hybrid gateway MANET protocol on the basis of dynamic TTL value adjustment. Aptikom J. Comput. Sci. Informat. Technol. **3**(2), 59–65 (2018)
23. K. Kishor, P. Nand, P. Agarwal, Secure and efficient subnet routing protocol for MANET. Execut. Ed. **9**(12), 200 (2018)
24. S.P. Yadav, D.P. Mahato, N.T.D. Linh, *Distributed artificial intelligence: a modern approach*, 1st edn. (CRC Press, 2020). https://doi.org/10.1201/9781003038467

25. S.P. Yadav, Emotion recognition model based on facial expressions. Multimed. Tools Appl. (2021). https://doi.org/10.1007/s11042-021-10962-5
26. Z. Zhang, Y. Xiao, Z. Ma, M. Xiao, Z. Ding, X. Lei, G.K. Karagiannidis, P. Fan, 6g wireless networks: Vision, requirements, architecture, and key technologies. IEEE Veh. Technol. Mag. **14**(3), 28–41 (2019)
27. S. Nayak, R. Patgiri, 6g communication technology: A vision on intelligent healthcare. arXiv preprint arXiv, 2005.07532 (2020)
28. M. Giordani, M. Polese, M. Mezzavilla, S. Rangan, M. Zorzi, To- ward 6g networks: Use cases and technologies. IEEE Commun. Mag. **58**(3), 55–61 (2020)
29. Y. Liu, J.J.Q. Yu, J. Kang, D. Niyato, S. Zhang, Privacy-preserving traffic flow prediction: a federated learning approach. IEEE Internet Things J., 1–1 (2020)
30. L.U. Khan, N.H. Tran, S.R. Pandey, W. Saad, Z. Han, M.N. Nguyen, C.S. Hong, Federated learning for edge networks: Resource optimization and incentive mechanism. arXiv preprint arXiv, 1911.05642 (2019)
31. J. Kang, Z. Xiong, D. Niyato, H. Yu, Y.-C. Liang, D.I. Kim, Incentive design for efficient federated learning in mobile networks: A contract theory approach, in *2019 IEEE VTS Asia Pacific Wireless Communications Symposium (APWCS)*, (IEEE, 2019), pp. 1–5
32. E. Jeong, S. Oh, H. Kim, J. Park, M. Bennis, S.-L. Kim, Communication-efficient on-device machine learning: Federated dis- tillation and augmentation under non-iid private data. arXiv preprint arXiv, 1811.11479 (2018)
33. N.H. Tran, W. Bao, A. Zomaya, N.M. NH, C.S. Hong, Federated learning over wireless networks: Optimization model design and analysis, in *IEEE INFOCOM 2019-IEEE Conference on Computer Communications*, (IEEE, 2019), pp. 1387–1395
34. K. Bonawitz, H. Eichner, W. Grieskamp, D. Huba, A. Ingerman, V. Ivanov, C.M. Kiddon, J. Konen, S. Mazzocchi, B. McMahan, T.V. Overveldt, D. Petrou, D. Ramage, J. Roselander, Towards federated learning at scale: System design, in *SysML 2019, 2019, to appear*. [Online]. Available: https://arxiv.org/abs/1902.01046
35. A.F. Atiya, A.G. Parlos, New results on recurrent network training: Unifying the algorithms and accelerating convergence. IEEE Trans. Neural Netw. **11**(3), 697–709 (2000)
36. Z. Wang, M. Song, Z. Zhang, Y. Song, Q. Wang, H. Qi, Beyond inferring class representatives: User-level privacy leakage from federated learning, in *IEEE INFOCOM 2019-IEEE Conference on Computer Communications*, (IEEE, 2019), pp. 2512–2520
37. L. Zhu, Z. Liu, S. Han, Deep leakage from gradients, in *Advances in neural information processing systems*, (2019), pp. 14 747–14 756
38. L. Li, H. Xiong, Z. Guo, J. Wang, C.-Z. Xu, Smartpc: Hierarchical pace control in real-time federated learning system, in *2019 IEEE Real- Time Systems Symposium (RTSS)*, (IEEE, 2019), pp. 406–418
39. A. Portnoy D. Hendler, Towards realistic byzantine robust federated learning. arXiv preprint arXiv, 2004.04986 (2020)
40. S. Guo, T. Zhang, X. Xie, L. Ma, T. Xiang, Y. Liu, Towards byzantine-resilient learning in decentralized systems. arXiv preprint arXiv, 2002.08569 (2020)
41. F. Ang, L. Chen, N. Zhao, Y. Chen, W. Wang, F.R. Yu, *Robust Federated Learning with Noisy Communication* (IEEE Transactions on Communications, 2020)
42. S.P. Yadav, K.K. Agrawal, B.S. Bhati, et al., Blockchain-based cryptocurrency regulation: an overview. Comput. Econ. (2020). https://doi.org/10.1007/s10614-020-10050
43. Y. Huang, Y. Su, S. Ravi, Z. Song, S. Arora, K. Li, Privacy-preserving learning via deep net pruning. arXiv preprint arXiv, 2003.01876 (2020)
44. T.-D. Cao, T. Truong-Huu, H. Tran, K. Tran, A federated learning framework for privacy-preserving and parallel training, arXiv preprint arXiv, 2001.09782 (2020)
45. Z. Jiang, A. Balu, C. Hegde, S. Sarkar, Collaborative deep learning in fixed topology networks, in *Advances in Neural Information Processing Systems*, (2017), pp. 5904–5914

46. L.U. Khan, N.H. Tran, S.R. Pandey, W. Saad, Z. Han, M.N. Nguyen, C.S. Hong, Federated learning for edge networks: resource optimization and incentive mechanism. arXiv preprint arXiv, 1911.05642 (2019)
47. J. Weng, J. Weng, J. Zhang, M. Li, Y. Zhang, W. Luo, *Deepchain: Auditable and Privacy-Preserving Deep Learning with Blockchain-Based Incentive* (IEEE Transactions on Dependable and Secure Computing, 2019)
48. Y. Zhan, P. Li, Z. Qu, D. Zeng, S. Guo, *A Learning-Based Incentive Mechanism for Federated Learning* (IEEE Internet of Things Journal, 2020)
49. H. Yu, Z. Liu, Y. Liu, T. Chen, M. Cong, X. Weng, D. Niyato, Q. Yang, A fairness-aware incentive scheme for federated learning, in *Proceedings of the AAAI/ACM Conference on AI, Ethics, and Society*,
50. A. Fallah, A. Mokhtari, A. Ozdaglar, Personalized federated learning: A meta-learning approach. arXiv preprint arXiv, 2002.07948 (2020)
51. Q. Wu, K. He, X. Chen, Personalized federated learning for intelligent IoT applications: a cloud-edge based framework. IEEE Open J. Comput. Soc., 1–1 (2020)
52. R. Hu, Y. Guo, H. Li, Q. Pei, Y. Gong, *Personalized Federated Learning with Differential Privacy* (IEEE Internet of Things Journal, 2020)
53. V. Kulkarni, M. Kulkarni, A. Pant, Survey of personalization techniques for federated learning. arXiv preprint arXiv, 2003.08673 (2020)

Federated Learning Using Tensor Flow

Tanu Solanki, Bipin Kumar Rai, and Shivani Sharma

1 Introduction

While the comparisons may be somewhat simplistic, the history of computing may be a decent proxy to what federated learning is all about. The large mainframe was used for doing the heavy computing in the earlier days. Gradually we moved toward the client server framework where the computation was distributed among the central server and the client computers. The federated learning architecture uses a similar model. Machine learning models were spread over the mobile devices instead of being computed on large, centralized machines. Due to the limitation of computational abilities of the mobile phones, the models were possible theoretically and not in the practical perspective, but due to the upgrade of mobile phones now, this sort of model is possible practically.

Large, varied, high-quality datasets were used for most of the AI algorithms. Huge amount of datasets are difficult to obtain specially in the field of healthcare. Federated learning opens the door for the AI algorithm to have knowledge from the huge range of data located at different sites. Thus, multiple organizations can collaborate with each other without sharing sensitive data with each other. The dataset developed with a wider range of data than any single organization may posses.

In their study, Sean Augestene et al. show that generative models prepared utilizing unified techniques and with formal differential protection ensures - can be utilized adequately to investigate numerous ordinarily happening information issues in any event, when the information can't be straightforwardly examined [1]. Andrew Hard et al.'s works exhibit the achievability and advantage of preparing language

T. Solanki
IT Department, Galgotia's College of Engineering and Technology, Uttar Pradesh, India

B. K. Rai (✉) · S. Sharma
IT Department, ABES Institute of Technology, Ghaziabad, Uttar Pradesh, India

S. P. Yadav et al. (eds.), *Federated Learning for IoT Applications*, EAI/Springer Innovations in Communication and Computing,
https://doi.org/10.1007/978-3-030-85559-8_10

models on customer gadgets without trading delicate client information to workers. The combined learning climate gives clients more noteworthy command over the utilization of their information and works on undertaking consolidating protection naturally with circulated preparing and collection across a populace of customer gadgets [2]. In Chaoyang He et al.'s work, they present FedML, an open exploration library and benchmark to encourage FL calculation advancement and reasonable execution correlation [3]. Oscar Li et al. disclose how to forestall the members' ground-truth names from conceivable spillage which isn't very much examined up to now [4]. Chaoyang He expresses that ascending up the convolutional neural network (CNN) size (e.g., width, profundity, and so forth) is known to adequately improve model precision. Even so, the large model size blocks preparing on asset cutting edge gadgets. For example, united learning (FL) may put excessive weight on the process ability of edge hubs, despite the fact that there is a solid functional requirement for FL because of its protection and privacy effects. FedGKT uses the substituting minimization method to deal with tiny CNNs training anxious hubs and periodically move their insight to a large worker-side CNN using information refining [5]. In their study, Peter Kairouz et al. state that FL epitomizes the standards of centered information assortment and minimization, and can relieve a large number of the foundational protection dangers and expenses coming about because of conventional, brought together AI and information science draws near. Kairouz et al. discussed the unstable development in federated learning research, as well as past advancement and recent open issues and difficulties [6].

Chaoyeng He et al. explain a focal worker free united learning calculation, named online push-sum (OPS) strategy, to deal with this difficult however nonexclusive situation [7]. In their work, Shaoxiong Ji et al. proposed mindful collection strategy limits the weighted distance between the worker model and customer models through iterative boundaries refreshing while goes to the distance between the worker model and customer models [8]. Sebastian Caldas et al. give a particular benchmarking system for learning in combined settings [9].

Daniel J. Beutel et al. proposed Flower, a FL structure which is both skeptic toward heterogeneous customer conditions and furthermore scales to countless customers, including versatile and implanted gadgets. Bloom's reflections let designers port existing versatile jobs with minimal overhead, paying little heed to the programming language or ML system utilized while likewise permitting analysts adaptability to explore different avenues concerning novel approaches towards advancing the cutting edge [10].

Tian Li et al. examine the interesting attributes and difficulties of united learning, give a wide outline of flow approaches, and blueprint a few bearings of future work that are pertinent to a wide scope of exploration networks [11]. Robin C Geyer et al. work on the expulsion of the weakness of convention to differential assaults, which could start from any gathering contributing during unified improvement. In such an assault, a customer's commitment during preparation and data about their informational index is uncovered through investigating the dispersed model. We tackle this issue and propose a calculation for customer-sided differential security safeguarding united advancement [12].

Tian Li et al. exhibited FedProx and considers more hearty assembly than FedAvg across a setup of practical combined datasets. Specifically, in profoundly heterogeneous settings, FedProx shows essentially more steady and exact assembly conduct comparative with FedAvg – improving outright test precision by 22% than normal [13]. Hongyi Wan et al. talked about experiments show that FedMA not only beats mainstream cutting-edge combined learning calculations on profound CNN and LSTM models prepared on genuine world datasets but also decreases the general correspondence trouble [14]. Hongyi Wan et al. perform tests which demonstrate that FedMA not just outflanks famous cutting-edge united learning calculations on profound CNN and LSTM models prepared on genuine world datasets but also decreases the general correspondence trouble [13]. Heiko Ludwig et al. clarified that federated learning (FL) is a way to deal with lead AI without bringing together preparing information in a solitary spot, for reasons of protection, privacy, or information volume. Notwithstanding, tackling combined AI issues raises issues far in excess of those of concentrated AI. Information researchers can plan and run unified learning occupations dependent on existing, brought-together AI models and can give undeniable level directions on the best way to run the alliance. The system applies to both deep neural networks just as "conventional" approaches for the most well-known AI libraries. It empowers information researchers to extend their extension from unified to united AI, limiting the expectation to learn and adapt at the start while likewise giving the adaptability to send to various process conditions and plan custom combination calculations [15]. For more clarification, the tensor flow documentation can be accessed [16].

In this chapter, we discuss how tensor flow is used for federated learning with the code implementation on the image dataset.

2 Federated Learning

Smarter, lower latency with less power consumption including privacy is allowed by federated learning for creation of the models. The real power of the mobile phone comes from the way in which one uses his/her phone, thus developing an improved model beyond simple updating and sharing. FL gives power to machine learning algorithms for obtaining experience from a huge amount of datasets located at distinct locations. Thus provides power to various organizations to collaborate with each other without sharing the confidential datasets, thus increasing the datasets for individual organization. The multiple repetitions at distinct places is used for training the model.

TensorFlow Federated (TFF) is an open-source framework for ML used on decentralized data and was invented for open research and experimentation with federated learning, where multiple clients across the globe with keeping their data locally train a shared model. The FL has been used widely for training prediction models for smart phones without uploading data to the servers.

TFF The TFF allows the developers to simulate the FL algorithms on models and datasets the developers can also experiment with the new algorithms. TFF's alliance is arranged in two layers.

Federated Learning (FL) API
In this layer, developers are allowed to implement the training and evaluation to already existent TensorFlow models on a high-level interface.

Federated Core (FC) API
The TensorFlow is combined with distributed communication operator within a strongly typed functional programming environment built on a lower-level interface.

2.1 Federated Learning (FL) API

These interfaces were designed with the intention that one can experiment with the federated learning without in detail knowledge of its working; also, FL algorithms can be evaluated and implemented on various existing models and data.

The following three parts are used in this interface as depicted in Fig. 1:

(a) *Models* – TFF uses a wrapped existing model done with the help of wrap classes and helper functions.
(b) *Federated Computation Builders* – There are a set of additional helping functions that were constructed for the federated computations for evaluation and training, with the use of existing models.
(c) *Datasets* – In Python programming language, the data can be downloaded and accessed for usage purposes in federated learning from Canned collections of data. It's always better to work with datasets that are available for downloading locally for developers who are new to this decentralized approach; thus, this data can be downloaded easily as compared to the decentralized datasets which

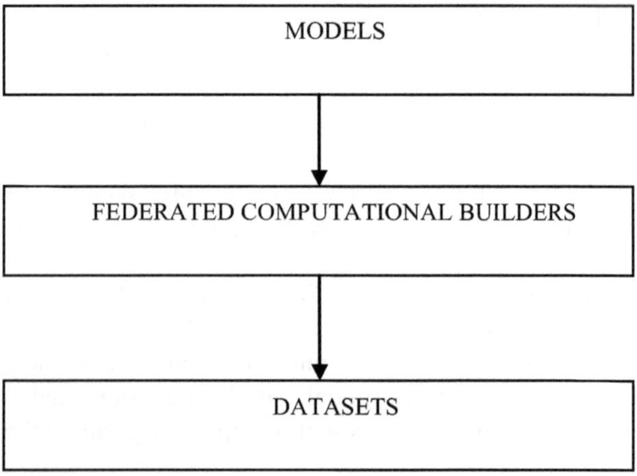

Fig. 1 Workflow of federated learning (FL) API

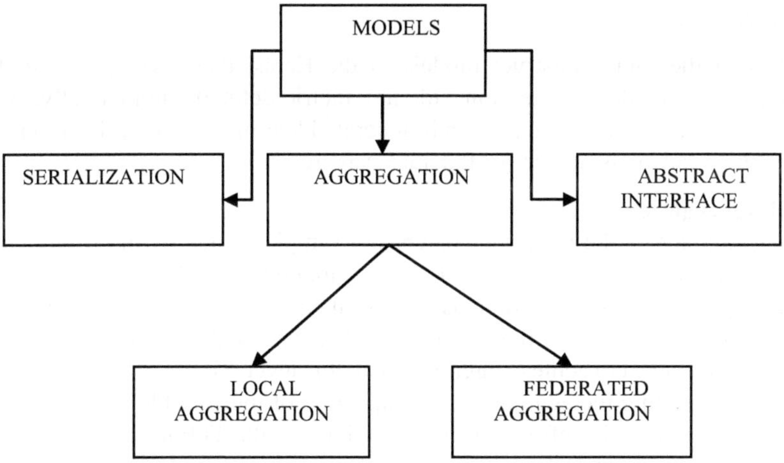

Fig. 2 Structural model analysis

can't be simply downloaded. The interface is defined in Tff.learning namespace; for research and simulation, tff.simulation is used. A runtime environment and a lower-level interface are offered by federated core (FC).

(a) *Models*

Model creation is done as the first step for the federated learning API which is further divided into three subparts, serialization, aggregation, and abstract interface; further aggregation is divided into local and federated aggregation as shown in Fig. 2.

(i) *Serialization*

A distributed learning scenario is supported by TFF where on a large number of heterogeneous clients, the machine learning code is executed. This is used for the support purpose of the mobile phone and the embedded devices which have limited resources. Sometimes, these devices are not capable for the python runtime hosting environment but may be competent enough for hosting a local TensorFlow runtime. The foundational assumption is made as model code is serializable as the TensorFlow graph. Recently, TensorFlow doesn't completely assist serializing and deserializing eager-mode TensorFlow. TF 1.0 pattern is currently followed for serialization in TFF, where a complete set of code is constructed inside tf.Graph controlled by TF. TFF can't work on a previously established model; thus, the concept for model definition is defined within the no-arg function that returns a tff.learning model, later called by TFF for ensuring all parts of the model are serializable. The metadata is required additional to all the things that is have.

(ii) *Aggregation*

Most of the users construct models in the Keras; these wrappers handle the aggregation of model update alongside any metric defined automatically. A minimum of two surfaces of aggregation in federated learning are used: local on-device aggregation and cross-device (or federated) aggregation.

Local Aggregation
Aggregation across the numerous batches of sample owned by the single client is referred by this level of aggregation. This is applied on both the model variables, continuing to sequentially evolve as local trained model, also the statistics, model will be updated locally as it iterates over single client local data stream. TensorFlow construction is used to achieve aggregation at this level. Firstly tf.Variables are constructed for holding the aggregates. The updation of the variable which holds various aggregates as a side effect, this happened due to the TFF invocation of forward pass method on the model many times. TFF invokes report_local_outputs method for compiling the summary statistics into the compact set for the export to the client.

Federated Aggregation
This aggregation works for the multiple-spread clients (devices) across the complete system. The output of local aggregation can be applied to the model variables which is averaged across clients, measures the model which can be further exported. The accountability for the aggregation performance is given to TFF. But if someone is the model creator, then he/she can control the process by following the given steps.

In Fig. 3, the general structure for controlling the process is defined and further elaborated in the below steps:

- The subset of clients that would participate in the evaluation and training purpose receives the initial model and also any parameter that has been required by the server.
- The model code has been invoked monotonously for producing the new set of parameters and metrics on the individual single client independently and in par-

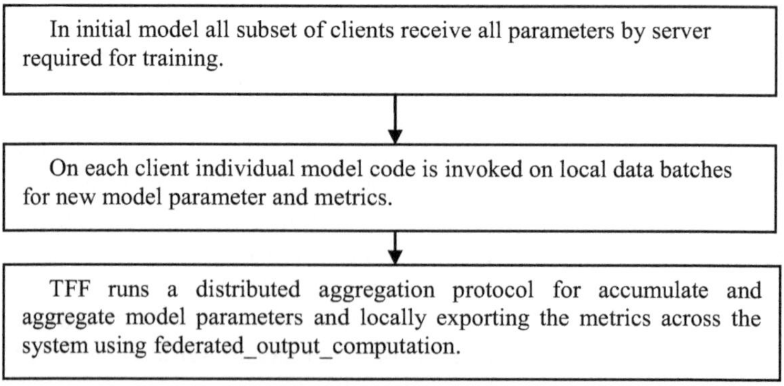

Fig. 3 Structure for controlling the process for model creator

allel mode as described in local aggregation. The model parameters and metrics that are locally exported over the system have been accumulated and aggregated by TFF using a distributed aggregation protocol.

- For aggregating the model parameters and locally exporting the various metrics throughout the system, a distributed aggregation protocol is run by TFF. The federated computation language is used expression of the logic in declarative manner in federated_output_computation.

(iii) *Abstract Interfaces*

The tff.learning.Model is represented by the basic constructor+metadata interface in the following manner:

- In order to report, we need model variables, forward pass and statistics was constructed by constructor, forward_pass and report_local_outputs functions. The Tensorflow constructed by these functions is serializable.
- The trainable, non-trainable, and local variables represented as metadata were returned in the form of subset by input_spec property, as well as three other properties. The TFF uses the information to regulate how to connect model parts to the federated optimization algorithms, and defining internal signature for verifying the correctness of the constructed system(checking the model is instantiated with the desired data for which it is constructed).

In addition to this, the federated_output_computation and report_local_outputs by the abstract interface tff.learning Model are used for the motivation of aggregating summary statistics.

(b) *Federated Computation Builders*

These computations to expand in the future where tff.learning provide several builders for the tff.Computations which are performing learning related tasks.

Architectural Assumptions for the Federated Computational Builders

The architectural assumption for the federated computational builder is depicted in Fig. 4.

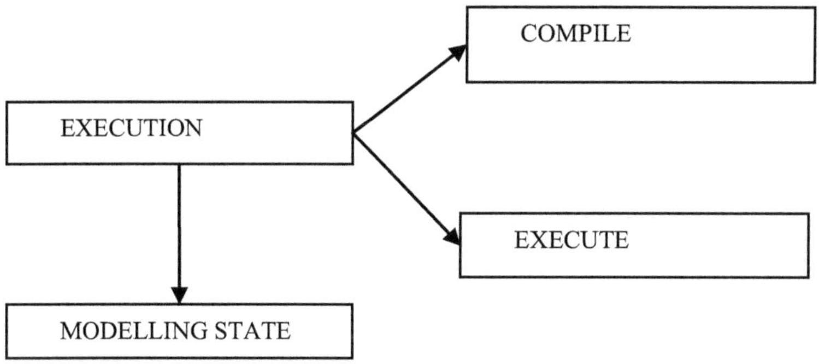

Fig. 4 Architecture model for the federated computation builders

Execution

The two definite phases while running a federated computation are as follows:

Compile: In this first step, TFF compile the federated learning algorithms into abstract serialized representation of the complete distributed computation. Other transformations can occur for supporting more efficient execution.

Execute: Provides the ways for the execution of the computation. Using a notebook, the execution is performed.

TFF's federated learning API generates federated computations, such as a training algorithm (using federated model averaging), or a federated evaluation (which includes a number of elements); most remarkable are:

- Model Serialized form + TensorFlow code constructed using federated learning framework for deriving models training and evaluation loop.
- In addition to the client and server in between communications declarative specification as well as gives the answer how this disperse communication is interleaved for TensorFlow code execution for client local and server local.

For the implementation and use of federated learning API, you need not worry with the detailed representation; a platform independent internal language which is different from Python is used for the representation of the federated computations. In Python code, tff.Computations are objects treated as Python callables. To invoke federated computation, just use python function which can be executed in local. This was used so that federated computations can be easily deployed on the android machines.

Modeling State

Even though there are many processes which are interesting in federated learning is stateful though TFF comes from a functional programming environment. Utilizing multiple rounds of federated model averaged by a training loop. In this procedure, the individual state which is evolving from one round to another and so on includes a set of model parameters which have been trained, possibly additional states related with the optimizer.

The current state acts as input, and the updated state acts as output accepted by computations; these processes are modeled by TFF as TFF is functional and stateful. For clarity of the stateful process, we need to find out from where the initial state is coming thus captured using helper class tff.templates.

Available Builders

Currently two builder functions for generating the federated computations for federated training and evaluation have been provided by TFF:

tff.learning.build_federated_averaging_process
tff.learning.build_federated_evaluation

(c) *Datasets*

At any instant of time, a small portion of active and available client devices are available for training from a huge population of about hundreds of millions of clients is depicted in the federated learning scenario. Thus, developer doesn't have any control over the number of clients participating in the process of training and evaluation. The typical round of training or evaluation only a small number of available clients, sampled randomly and furthermore it is also impractical to coordinate with millions of clients.

As all processing is expressed as an aggregate operations on an abstract group of unknown client, group might vary from one circle of training to another circle of training, by design of the federated computation, expressed to the exact set of participants. The unbreakability of the computation to the concrete contributors, the concrete data is inputted to the computation is modelled outside the computation itself.

3 Federated Core

Federated core (FC) is understood as a programming environment for the computation of the distributed system, where computations involve multiple computers (mobile phone, tablet, sensors, etc.) and where individual devices perform nontrivial processing locally and communicate with each other on the network for the development of the work.

Here, we use a less generic term which is known as federated computation for describing the various types of algorithms which can be defined in this framework because TFF does not target all available types of distributed algorithms as this word "distributed" is very common.

The Federated computation can be understood as the algorithms which are expressed using the pseudocode in the research publications which describe the new algorithm of distributed learning.

The work of the FC is to enable closeness compact representation, at a close psuedocode abstraction level, of the programing logic that is not a psuedocode, rather it is executable in a diversity of target environment.

The actions of the participant system, described in a collective manner then these must be expressed by FC are the basic characteristics that can be used for FC implementation. The individual devices locally modifying data, devices coordinating their work by centralized arranger for transmitting, collecting their outputs.

While TFF has been intended to have the option to go past basic client server structures, the idea of aggregate handling is central. This is because of the beginnings of TFF in unified learning, an innovation initially intended to help calculations on conceivably delicate information that stays leveled out of customer gadgets, and that may not be basically downloaded to a brought-together area for protection reasons. While every customer in such frameworks contributes information and preparing power toward figuring an outcome by the framework (an outcome that we would by and large hope to be of worth to every one of the members), we additionally endeavor at saving every customer's security and namelessness.

In this manner, while most structures for appropriated figuring are intended to communicate preparing from the viewpoint of individual members, that is, at the degree of individual highlight point message trades, and the relationship of the member's neighborhood state changes with approaching and cordial messages, TFF's Federated core is intended to portray the conduct of the framework from the worldwide framework broad view.

The distributed framework, for general purposes, offer operations send and receive as a building blocks, FC providing building blocks as tff.federated_sum, tff.federated_reduce, or tff.federated_broadcast that encapsulate simple distributed rules.

4 Conclusion

The Federated Learning makes smart phones learn collaboratively through a mutual prediction model while keeping all data that is used for training on device, disconnecting the capacity to do machine learning forms the demand to store the data in the cloud. The step set has been defined in the tensor flow documentation for the connection of the tensor flow to the federated learning. Federated learning is very important for the security and privacy of the dataset which is a sudden concern for all. Machine learning has been used extensively for federated learning using tensor flow. Federated learning API and federated core API have been discussed here with the functions that are required for their implementation.

References

1. S. Augenstein, H.B. McMahan, D. Ramage, S. Ramaswamy, P. Kairouz, M. Chen, R. Mathews, Generative models for effective ML on private, decentralized datasets. arXiv preprint arXiv:1911.06679 (2019)
2. A. Hard, K. Rao, R. Mathews, S. Ramaswamy, F. Beaufays, S. Augenstein, D. Ramage, Federated learning for mobile keyboard prediction. arXiv preprint arXiv:1811.03604 (2018)
3. C. He, S. Li, J. So, M. Zhang, H. Wang, X. Wang, ... S. Avestimehr, Fedml: A research library and benchmark for federated machine learning. arXiv preprint arXiv:2007.13518 (2020)
4. O. Li, J. Sun, X. Yang, W. Gao, H. Zhang, J. Xie, ... & C. Wang, Label leakage and protection in two-party split learning. arXiv preprint arXiv:2102.08504 (2021)
5. S.P. Yadav, S. Yadav, Fusion of medical images using a wavelet methodology: A survey. IEIE Transact. Smart Proces. Comput. **8**(4), 265–271 (2019). https://doi.org/10.5573/IEIESPC.2019.8.4.265
6. P. Kairouz, H.B. McMahan, B. Avent, A. Bellet, M. Bennis, A.N. Bhagoji, S. Zhao, Advances and open problems in federated learning. arXiv preprint arXiv:1912.04977 (2019)

7. C. He, C. Tan, H. Tang, S. Qiu, J. Liu, Central server free federated learning over single-sided trust social networks. arXiv preprint arXiv:1910.04956(2019)
8. S. Ji, S. Pan, G. Long, X. Li, J. Jiang, Z. Huang, Learning private neural language modeling with attentive aggregation. In 2019 International Joint Conference on Neural Networks (IJCNN) (2019, July, pp. 1–8). IEEE
9. S.P. Yadav, K.K. Agrawal, B.S. Bhati et al., Blockchain-based cryptocurrency regulation: An overview. Comput Econ. (2020). https://doi.org/10.1007/s10614-020-10050-0
10. D.J. Beutel, T. Topal, A. Mathur, X. Qiu, T. Parcollet, N.D. Lane, Flower: A friendly federated learning research framework. arXiv preprint arXiv:2007.14390 (2020)
11. T. Li, A.K. Sahu, A. Talwalkar, V. Smith, Federated learning: Challenges, methods, and future directions. IEEE Signal Process. Mag. **37**(3), 50–60 (2020)
12. T. Li, A.K. Sahu, M. Zaheer, M. Sanjabi, A. Talwalkar, & V. Smith, Federated optimization in heterogeneous networks. arXiv preprint arXiv:1812.06127 (2018)
13. H. Wang, M. Yurochkin, Y. Sun, D. Papailiopoulos, Y. Khazaeni, Federated learning with matched averaging. arXiv preprint arXiv:2002.06440 (2020)
14. S.P. Yadav, D.P. Mahato, N.T.D. Linh, *Distributed Artificial Intelligence: A Modern Approach*, 1st edn. (CRC Press, 2020). https://doi.org/10.1201/9781003038467
15. H. Ludwig, N. Baracaldo, G. Thomas, Y. Zhou, A. Anwar, S. Rajamoni, A. Abay, IBM federated learning: An enterprise framework (2020)
16. https://www.tensorflow.org/federated/federated_learning

Cyber Security and Privacy of Connected and Automated Vehicles (CAVs)-Based Federated Learning: Challenges, Opportunities, and Open Issues

Naziya Hussain, Preeti Rani, Harsha Chouhan, and Urvashi Sharma Gaur

1 Introduction

Federated learning (FL) is a machine learning (ML) procedure which allows an approach to be trained across many distributed entities while preserving privacy of their local information. The method allows for issues like data privacy, protection, and access rights to be addressed. Despite the benefits, applying FL to new cellular mobile networks, such as CAVs-based intelligent transportation systems (ITS), presents major challenges [40].

Kuderer et al. [25] have already used machine learning techniques in different ways to enhance the autonomous vehicle performance: computer vision for detecting complications and ML for adjusting speed of the network. Because of the possible huge various self-driving vehicles and necessity for them to reply easily to real-world conditions, the outdated cloud solution poses a safety danger. FL may be a solution for regulating data transfer volume and speeding up autonomous vehicle learning processes. For such linked autonomous vehicles, the advantages include [43] better provision of smooth message transfer as well as low-latency Internet services while on the move.

In academia and business, the idea of a vehicular ad hoc network (VANET) is gaining traction. It's an essential part of the ITS that aims to enhance driving safety through vehicle contact or communication with roadside infrastructure [16, 42]. With the support of a roadside system and the network, this VANET foundation

N. Hussain (✉) · H. Chouhan · U. S. Gaur
School of Computers, IPS Academy, Indore, MP, India

P. Rani
Department of Electronics and Communication Engineering,
SRM Institute of Science and Technology, Delhi-NCR Campus,
Delhi-Meerut Road, Modinagar, Ghaziabad, UP, India

S. P. Yadav et al. (eds.), *Federated Learning for IoT Applications*, EAI/Springer Innovations in Communication and Computing,
https://doi.org/10.1007/978-3-030-85559-8_11

helps end users get early access to traffic incident info. VANET may be used as a monitoring and early warning system to provide a precise triggering signal to mitigation systems in the early stages of a potential traffic incident [33].

The rapid growth of cyberspace has been aided by software-defined networking (SDN), fog computing, as well as big data, examples of new networking and computing developments [28–30]. In the meantime, one of the most pressing topics in cyberspace is cyber security [15, 55]. Cyber security has had a huge effect on a variety of sensitive structures. Traditional protection relies on the stationary control built on special edges or nodes to handle network protection according to predefined rules, such as firewalls, intrusion detection, and prevention systems (Table 1).

Caching output can be improved by placing common caching contents at RSUs, which is primarily dependent on awareness of content popularity. Many factors affect the popularity of content, including vehicle users' demographic details (e.g., age and gender) and vehicle mobility patterns. To improve cache efficiency, the above information must be considered when forecasting determining which contents to cache in RSUs based on content popularity.

Table 1 Applications, prominent features, and designing goals of FL as well as other distributed approaches

Methods	Prominent features and design goals	Usages
Distributed learning (DL)	The aim is to deliver a general evaluation of the limitations under research The global model is not communicated to the local students	Wireless sensor network based on distributed learning (DL)
Parallel learning (PL)	The aim is to speed up the learning method and increase the approach's scalability Data is distributed in a unique way and not widely disseminated between the users	DL in the context of data centers
Ensemble learning (EL)	Data is disseminated in a IID manner, and there are no message constraints to consider The target is to create a best system by education from a mixture of various types of systems	Bagging, boosting, and stacking algorithms for face recognition, remote sensing, and other applications
Federated learning (FL)	The aim is to train the system by utilizing naturally distributed datasets through multiple users The global system is then fed back to the confined users for usage Data is disseminated in a non-IID manner and disseminated widely between local users In which communication system restrictions in accessing the data, such as privacy, bandwidth limitations, protection, and power	Edge computing as well as caching Self- driving vehicles Federated machine learning for variety management Coexistence of heterogeneous schemes (such as, DSRC, and c-V2X) 5G core network in FL

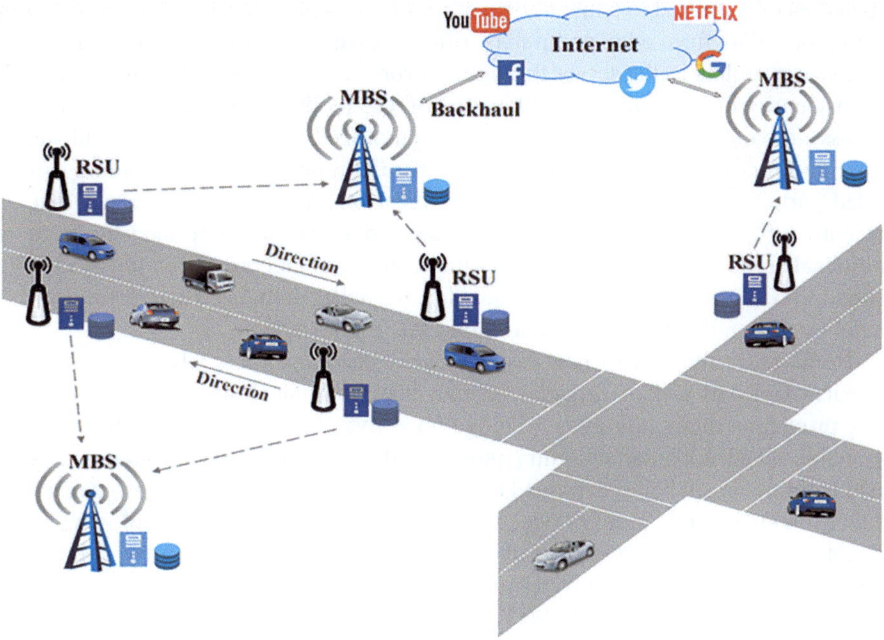

Fig. 1 Three-tier architecture of the connected vehicles on road side unit in FL

Yu et al. [61] use FL to strategize a mobility-aware proactive data caching system for connected vehicles to solve the dilemma. It is a three-tiered architecture, as shown in Fig. 1. The bottom sheet is made up of vehicles that need content. The middle layer is comprised of numerous RSUs with cache-enabled edge servers. There is a cache-enabled MBS in the top layer. In an RSU, several associated vehicles work together to train a common global learning model (LM). To begin, the RSU distributes a global model to connected vehicles.

2 Review of Federated Learning

Federated ML is a modern decentralized methodology that considers problems like anonymity and resource limits, as well as others. By decentralizing model training and storing data where it is generated, it allows use of on-device processing resources and previously unexploited private data. Niknam et al. [40] offer an open summary to the common perception of federated ML as an expansion to Google's latest federated approach [3, 38]. The main characteristics of FL that distinguish it from other decentralized learning (DL) methods are then discussed.

Through DL from the essential server to end devices and protecting consumer privacy, AI profits fields of confidential data as well as heterogeneity. There is insufficient data to stay centrally on the server side due to direct access restrictions on such data (as opposed to conventional ML). (2) Data privacy rights are protected by using local data from edge computers, often known as clients, rather than transmitting sensitive data to the server through network asynchronous communication. (3) Data security ensures that the AI benefits of ML models can be efficiently applied across domains. Instead of relying on a centralized server, iterative local model training processes on end devices often share computing resources among interested parties. FL has become one of the most rapidly expanding areas in ML in recent years, due to its decentralized data description and security and privacy capabilities that pledge to conform to emerging consumer data protection laws [35]. FL, in addition to privacy, provides ML benefits to smaller domains where there is insufficient training data to build a stand-alone ML framework.

However, in the face of new cyber security challenges like advanced persistent threats (APTs) as well as zero-day attacks, it's important to stay vigilant. This passive protection methodology is no longer effective [41]. Due to the proliferation of systemic attack tools, attack entry points, and high-level intrusion modes, a wide range of threats are being deployed to optimize the security level of core systems as cyber threats become more prevalent and long-lasting, the cost of deploying cyber threats is reduced by the variety of systemic attack tools, attack entry points, and high-level intrusion mode a wide range of threats in order to optimize the security level of core system properties [45]. To introduce modern in order to provide adaptive security management and regulation. The system gathers historical and current security data to provide adaptive security management and regulation to make informed decisions (Table 2).

FL was first developed by Google in 2016, and it is a procedure which is often utilized for collaborative testing of information data in various ED for centralized system processing and utilized in situations as per the input system optimization

Table 2 Federated learning threats

Threats severity	Severity	ML framework	Source of vulnerability
Poisoning	H	DML/FL	Client data manipulations (CDM) and compromised central server (CCS)
Inference	H	FL	CDM and CCS
Backdoor attacks	H	DML/FL	CDM and CCS
GANs	H	FL	CDM and CCS
Malicious server	H	DML/FL	CCS
Communication bottlenecks	H	DML/FL	Weaker communication bandwidth (WCB)
Free-riding	M	FL	Clients
Unavailability	M	FL	Clients
Eavesdropping	M	FL	Clients
Interplay along data protection laws	L	FL	Weaker communication protocol (WCP)
System disruption IT downtime	L	FL	FL environment implementer

[59, 60]. FL is reshaping the financial, medical, and urban security fields as a new wave of artificial intelligence (AI) technologies [31].

By overcoming the core challenges of commercial AI technology bottlenecks, such as data privacy and data scarcity, it is reshaping the banking, medical, and urban security fields.

FL is built on the concepts of concentrated data gathering and minimization, and it will help alleviate much of the structural privacy issues and costs that come with traditional, organized ML and data science approaches. Kairouz et al. [24], motivated by the exponential growth in FL science, address recent developments and present a comprehensive array of open problems and challenges.

Xu et al. [59] look at two major problems in the deep neural network (DNN) training process: (1) how to defend the privacy of users during the training phase (i.e., local gradients) and (2) how to maintain the honesty of the combined information data resumed from the server during the training phase (i.e., local gradients) and how to defend the privacy of users during the training phase (i.e., local gradients). To solve the problems mentioned above, several approaches based on the secure privacy-preserving FL have been proposed and tested in a variety of scenarios. Allowing clients to verify if the cloud server is up and running while maintaining user privacy during the training phase, on the other hand, is still in the works.

Since FL allows for joint ML model training as well as DL for mobile edge (ME) system optimization, it can be used as an enabler in mobile edge networks. However, heterogeneous devices with varying constraints are complex in the large-scale as well as composite ME network. This advances questions about bandwidth prices, resource sharing, and privacy and protection as FL scales up. Lim et al. [32] provide an overview of FL's history and fundamentals. Then we go over the aforementioned FL implementation challenges and existing solutions. In addition, authors show how FL can be used to optimize mobile edge networks and, finally, go through some of FL's major obstacles and potential research directions (Table 3).

As compared to traditional optimization methodologies and current deep learning techniques, we suggest combining deep reinforcement learning (DRL) system and the FL platform with ME systems to optimize ME computing, caching, and communication in order to bring more information to the edge systems.

Wang et al. [54] develop the "In-Edge AI" context to intelligently use association among devices as well as edge nodes to share learning parameters for better model training and inference, allowing them to perform dynamic system-level optimization and application-level enhancement while reducing unnecessary system communication load.

The three main components of FL are data sources, FL systems, and users. Figure 2 depicts the three's relationship. We impose distinction privacy on the mined features and design a modified normalization system to defend the privacy of customers while also improving test accuracy. Zhao et al. [62] establish that our normalization system outdoes batch normalization, while sorts are subject to variance privacy defense. Create an incentive scheme to reward crowdsourcing FL challenge participants for enticing more consumers to participate.

Table 3 FL defense approaches

Defenses	Explanation	Attacks	Authors
Sniper	Configure Euclidean distance form based on global server to eliminate adversarial informs	Poisoning	Cao et al. [9]
Knowledge distillation	Transmitting data by fully trained system to another model	Eavesdropping Inference GANs	Li & Wang [26]
Anomaly Detection	Observing for apprehensive updates of users	Poisoning trojans Model update poisoning	Blanchard et al. [5], Xie et al. [57], Ito et al. [21]
Moving target defense	Obfuscating source of vulnerability	Eavesdropping	Colbaugh & Glass [10]
Pruning	Reduce the size of neural network (NN) model	Backdoor attacks, system calculation of the communication costs	Jiang et al. [23]
Data sanitization	Remove or delete the data just after use	Poisoning attacks	Cretu et al. [61]
Trusted execution environment	Delivers integrity as well as confidentiality of the code implementation on a server	Malicious server	Mo, & Haddadi [36]
Fool's gold	As per variety of client information, Sybil attacks are well-known	Sybil-based label flipping backdoor poisoning attacks	Fung et al. [14]
Federated multitask learning	Train models by various correlated tasks simultaneously	Device drop fault tolerance	Smith et al. [47]

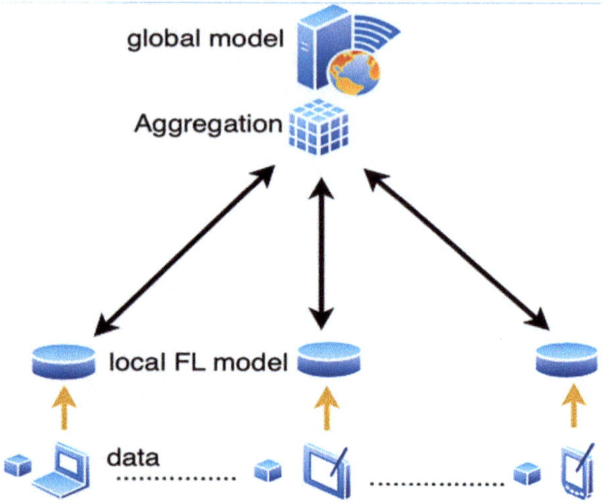

Fig. 2 Basic federated learning architecture

A solution based on FL-enabled intelligent F-RANs is projected as a viable option that can fully use fog computing as well as AI. The basic concept of model compression and accuracy loss correction is studied; some understanding into the essence of FL in F-RANs is gained. Some main techniques are implemented to completely incorporate the communication, computation, and storage capacity of F-RANs which are used to make FL easier to incorporate. Future advances in FL-enabled intelligent F-RANs, including potential implementations and open problems, are also explored [57].

FL effectively addresses the issue of data islands by allowing two or more data-using organizations (clients) to share data without having to contribute data. When the national model of FL and centralized data management are matched, the global model of FL will have the same modelling effect [7].

3 Cyber Security in Vehicular Network

Chowdhury, M., & Wang, K. C. [6], presents the various wireless technologies have been utilized to livelihood the data transmission necessities of various ITS applications. The accessibility and viability of wired and wireless connectivity solutions (Wi-Fi, WiMAX, LTE), as well as the data transmission requirements of specific applications, influence the choice of wireless communication option [6, 13]. Though current ITS implementations are infrastructure-based, high-speed wireless connectivity among moving vehicles as well as transportation infrastructures is the next big implementation of wireless technology between the transportation grid (V2I and V2V). Fast achievement, fast message transmission (i.e., low latency) with high reliability, and the highest security and privacy standards are needed for V2I as well as V2V communication-enabled CVT safety submissions [47].

For a stable VANET network, a novel attack-resilient and efficient QoS-based GPCR-ARE protocol solution is used. Vehicle protection can be improved using a deep neural network-based intrusion detection system (IDS) (DNN). The DNN system is trained using feature vectors extracted from VANET system data that include the likelihood of each class perceptive to regular and attack packets, allowing the sensor to identify any attack vehicle. In comparison to the traditional artificial neural network (ANN) architecture used in the IDS scheme, the enhanced methodology takes into account recent advances in ANN research, such as modifying the initial parameters of deep belief networks in unsupervised settings (DBN) as well as an attack-resistant and efficient system, improving detection accuracy [20, 46].

However, this contact and the other assets involved are vulnerable to significant threats, allowing attackers to initiate a variety of attacks and breach the protection and privacy of vehicle users. Ahmad et al. [1] examine cyber security and suggest an asset-based approach to security in the VANET. To begin, first, it locates relevant assets in the VANET. Second, it offers a comprehensive taxonomy of vulnerabilities and risks to these properties, and third, it classifies and objectively evaluates potential VANET attacks.

The aim of Rizvi et al. [46], by fixing previous security susceptibilities and threats, creates a threat system for CAR networks. We also show how to keep a CAR network from both internal and external networks using a distributed firewall architecture to overcome these security challenges.

Modern vehicular systems have a number of advantages: vehicle networks reduce traffic congestion while also improving safety and fuel economy. Many industry experts, on the other hand, believe that the most important part of a vehicle, the Controller Area Network (CAN) bus, is safe and can't be simply hacked; additionally, computer security authorities have little experience with the CAN bus.

Pan et al. [60] gives an in-depth overview of the CAN bus and possible situations in which a vehicle's CAN bus is compromised. Furthermore, malicious attackers can save money by using CAN vehicle attacking in mobile malware, but vehicle engineers will find it difficult to moderate the security issues.

Cyberattacks on today's vehicles that threaten vehicular communications are a very real possibility. El-Rewini et al. [12] propose a three-layer system to better understand automotive security threats (sensing, contact, and control). Vehicle dynamics and environmental sensors make up the sensing layer; eavesdropping, jamming, and spoofing attacks are all possible. A vehicle dynamics sensor and environmental sensors make up the sensing layer, while eavesdropping, jamming, and spoofing attacks can occur at all communications layers, including in-vehicle and V2X communications. The control layer, which is at the top of the hierarchy, enables autonomous vehicular functionality, such as pace, acceleration, and steering. Attacks on the coordination and sensing layers will spread upward, affecting functionality and compromising the control layer's protection. It offers a current state-of-the-art analysis of communication layer attacks and threats, as well as countermeasures.

Vehicle-to-Everything (V2X) connectivity will be added to the vehicular network in the near future, allowing vehicles to communicate with any smart devices on the roadside. Due to heterogeneous nodes, varying speeds, and unreliable connections, the vehicular network poses numerous challenges, and conventional protection measures are not always efficient. As a result, a wide range of research projects have been undertaken to improve security solutions while taking network specifications into account. They are existing systematic literature and taxonomy of current security clarifications for V2X communication technologies in [2] present complete literature and taxonomy of emerging security strategies for V2X connectivity technologies. Include debates and comparisons concerning specific applicable criteria. Include debates and comparisons with respect to certain applicable criteria. Also included should be a threat review for V2X enablement technologies.

4 Federated Learning (FL) Algorithm Design

Local and global computing tools, as well as communication resources, are used by FL. Long- and short-term memory, convolutional neural networks (CNN), and Naive Bayes (NB) schemes are examples of ML techniques. Non-convex FL models can be trained using a variety of optimization schemes, including federated averaging (FedAvg) and FedProx [39]. FedProx is a tweaked variant of FedAvg that takes into account numerical and system variability on the end device. FedAvg generates local model weights on a group of devices using stochastic gradient descent (SGD).

Following that, the BS edge computing server computes an average of the local weights. The distinction between FedProx and FedAvg is in the local system minimizing the objective functionality, which takes into account FedAvg's objective function plus a supplementary proximal term. In this way, FedProx mitigates the impression of non-independent and identically distributed (non-IID) machine data on the global learning model (LM). FedAvg cannot ensure hypothetical convergence, but FedProx should show it in Fig. 3.

Despite the discrepancies in system capabilities, in federated averaging in the global FL model computations of FedProx and FedProx, both instruments are weighted similarly (e.g., hardware). Fairness allowed FedAvg algorithm [26] was proposed to capture such fairness among devices. By altering the objective feature of the standard FedAvg algorithm, Fairness allowed devices with low performance are given a higher weight by FedAvg. Local devices with a high empirical damage are given more relative weight in the fairness analysis allowed FedAvg to implement possible fairness and minimize preparation correctness adjustment. Local and

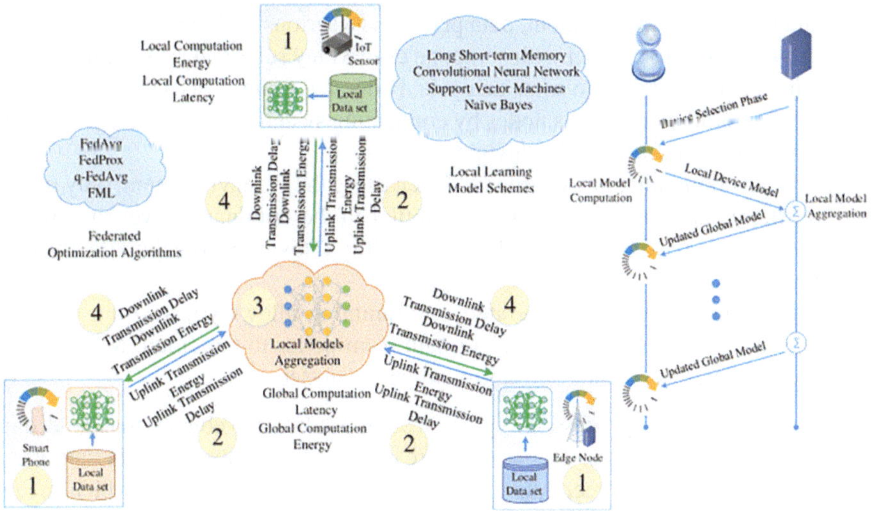

Fig. 3 Federated learning (FL) algorithm design

global computing tools, as well as communication resources, are used by FL. Every local system can employ a diversity of ML techniques, including NBs systems, long- and short-term memory, and CNNs. Non-convex FL models can be trained using a variety of optimization schemes, including federated averaging (FedAvg) and FedProx [39]. FedAvg has been changed to build FedProx that accounts for end-device statistical and machine heterogeneity. FedAvg uses stochastic gradient descent (SGD) to generate local model weights on a group of devices. By altering the architecture of the so-called communication-efficient distributed dual coordinate ascent (CoCoA), Smith et al. [53] suggested a federated multitask learning approach. CoCoA supports linear regularized loss minimization objectives to allow a broad range of ML models [34]. CoCoA effectively combines partial results from local computations using a primal-dual form to solve optimization problems. Each round of CoCoA uses distributed optimization to deal with machine- and statistical heterogeneity, enabling any arbitrary optimization algorithm on a local dataset to solve a local learning problem, use any arbitrary optimization algorithm on a local dataset.

5 Federated Machine Learning (ML) Method Layout

McMahan et al. [37] Google enhanced a novel training scheme for learning collaboratively from different devices. This technique, known as FL, allows the model to be jointly trained on multiple nodes without the need for data collection. The net advantage of this method is the protection of privacy and the ability to learn from a large number of nodes. Because of the following factors, this training method is advantageous:

1. Data privacy: At this moment, data privacy is a serious problem. Users are afraid to share their information with someone else. Many rules and regulations tend to gather data in order to do some analysis. Federated training, as a solution, allows for data training on various nodes by connecting system to data sources. The data from nodes is never seen by anyone.
2. Reduced communication overhead: When using the FL environment, only model constraints and inclines can be transmitted to the dominant server. It can also select clients based on bandwidth accessibility. This improves the effectiveness of touch [11].
3. Train on large datasets: Federated programming allows you to train on large datasets without having to transmit data from nodes to the cloud. Model accuracy improves as a result of training on large datasets.

As per Fig. 4, FL process is typically composed of K vehicles that hold data samples D_i locally, where $|D_i| = l_i$. There are a total number of data samples I s $\sum_i l_i = l$. Each vehicle keeps its data private, i.e., $D_i = \{x_1^i, \ldots, x_{l_i}^i\}$, which means that it is not shared with the chief S or any other vehicle $i \neq j$, $\forall j \in k$. The chief trains a model f with $W_G \in R^n$ as global parameter vector, where n is the dimension

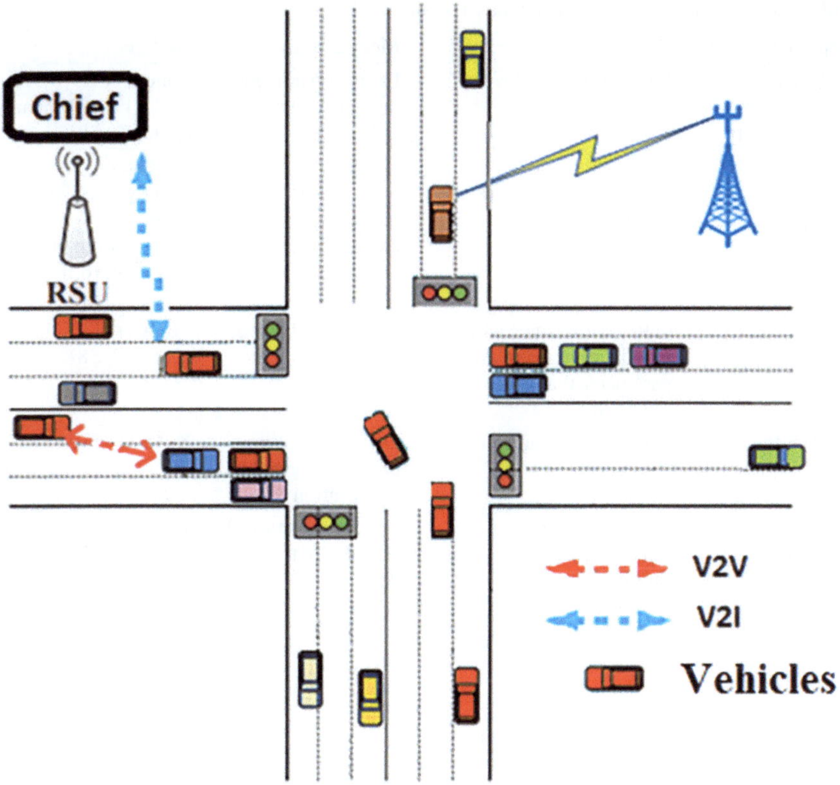

Fig. 4 Communication model for vehicles using V2I with the chief implemented at the RSU

of the parameter space. The global model that results is obtained through distributed training (DL) and aggregation across the K vehicles. The goal of the training is to generalize beyond the scope of test dataset D_{test}. The chief selects a random subset of k vehicles at each time step t. Every vehicle $i \in |k|$ develops a local system to minimize a loss function over its own data D_i. The vehicle begins the local training from the global model W_G^t received from the chief and runs an approach such as the stochastic gradient descent for a number of ε epochs with a batch size of B. After that, each vehicle then obtains a local weight vector W_G^{t+1} and then computes its local model update $\delta_i^{t+1} = W_i^{t+1} - W_G^t$. The local model update is sent back to the chief. To obtain the global model update W_G^{t+1} for the next iteration, an aggregation mechanism is used: $W_G^{t+1} = W_G^{t+1} + \sum_{i \in |k|} \alpha_i \delta_i^{t+1}$ where αi = α_i = l_i/l and $\sum_i \alpha_i = 1$ (Table 4).

Table 4 Enhancement privacy preservation method in FL

Approach	Cost	Methodology	Authors
Multi-party computation of confidentiality	Encryption reduces performance	Encrypt parameters that have been uploaded	Canetti et al. [8], Hao et al. [17, 18], Zhang et al. [57, 58]
Privacy differential	Increased noise in the client's model causes accuracy loss	Toss in any random noise to the parameters you've already entered	Jajodia, & van van Tilborg [22], Zhu & Philip [59]; Bagdasaryan et al. [4], Wei et al. [62]
Hybridized	Costs are low in terms of both performance and precision. Encrypt the parameter that was modified	Encrypt the parameter that was modified	Hao et al. [17, 18], Truex et al. [52], Zhang et al. [57, 58]
VerifyNet	Communication overhead	Double-masking protocol Verifiable aggregation results	Xu et al. [59]
Adversarial training	Computation power, training time for adversarial samples	Include adversarial samples in training data	Tramèr et al. [51], Hayes & Ohrimenko [19], Zhang et al. [57, 58]

6 Conclusion

This chapter addresses the gaps of existing research by examining the factors that influence user adoption of CAVs when it comes to cyber security and privacy concerns. Consumers, industry, and governments all over the world are fascinated by CAVs. While CAVs are designed with capabilities and features that have the potential to improve protection, satisfaction, comfort, and convenience, CAVs also get new security and privacy risks, posing a mix of physical and digital threats. This attacks model establishes that FL is exposed to mobile attackers developing the medium to perform model poisoning in its standard form. Validating robustness to attackers of the type considered in this chapter is yet to be accomplished. In future work, we want to investigate sophisticated security mechanisms that can provide assurances against CAV attackers in the future. In particular, encryption, localization, behavioral analysis, and clustering could be promising detection mechanisms in this situation.

References

1. F. Ahmad, A. Adnane, V.N. Franqueira, A systematic approach for cyber security in vehicular networks (2016)
2. A. Alnasser, H. Sun, J. Jiang, Cyber security challenges and solutions for V2X communications: A survey. Comput. Netw. **151**, 52–67 (2019)

3. K. Bonawitz, H. Eichner, W. Grieskamp, D. Huba, A. Ingerman, V. Ivanov, ... , J. Roselander, Towards federated learning at scale: System design. arXiv preprint arXiv:1902.01046 (2019)
4. E. Bagdasaryan, O. Poursaeed, V. Shmatikov, Differential privacy has disparate impact on model accuracy. Adv. Neural Inf. Proces. Syst. **32**, 15479–15488 (2019)
5. P. Blanchard, E.M. El Mhamdi, R. Guerraoui, J. Stainer, Machine learning with adversaries: Byzantine tolerant gradient descent, in *Proceedings of the 31st International Conference on Neural Information Processing Systems* (2017, December), pp. 118–128
6. M. Chowdhury, K.C. Wang, Distributed intelligent traffic sensor networks, in *Transport Science and Technology*, (Emerald Group Publishing Limited, 2006)
7. Z. Chai, H. Fayyaz, Z. Fayyaz, A. Anwar, Y. Zhou, N. Baracaldo, ..., Y. Cheng, Towards taming the resource and data heterogeneity in federated learning, in *2019 {USENIX} Conference on Operational Machine Learning (OpML 19)* (2019), pp. 19–21
8. R. Canetti, U. Feige, O. Goldreich, M. Naor, Adaptively secure multi-party computation, in *Proceedings of the Twenty-Eighth Annual ACM Symposium on Theory of Computing* (1996, July), pp. 639–648
9. D. Cao, S. Chang, Z. Lin, G. Liu, D. Sun, Understanding distributed poisoning attack in federated learning, in *2019 IEEE 25th International Conference on Parallel and Distributed Systems (ICPADS)*. IEEE (2019, December), pp. 233–239
10. R. Colbaugh, K. Glass, Moving target defense for adaptive adversaries, in *2013 IEEE International Conference on Intelligence and Security Informatics*. IEEE (2013, June), pp. 50–55
11. Du, Zhaoyang, Celimuge Wu, Tsutomu Yoshinaga, Kok-Lim Alvin Yau, Yusheng Ji, and Jie Li. "Federated learning for vehicular internet of things: Recent advances and open issues." IEEE Open Journal of the Computer Society 1 (2020): 45–61
12. Z. El-Rewini, K. Sadatsharan, D.F. Selvaraj, S.J. Plathottam, P. Ranganathan, Cybersecurity challenges in vehicular communications. Vehicular Commun. **23**, 100214 (2020)
13. K.C. Dey, A. Mishra, M. Chowdhury, Potential of intelligent transportation systems in mitigating adverse weather impacts on road mobility: A review. IEEE Trans. Intell. Transp. Syst. **16**(3), 1107–1119 (2014)
14. C. Fung, C.J. Yoon, I. Beschastnikh, Mitigating sybils in federated learning poisoning. arXiv preprint arXiv:1808.04866 (2018)
15. Z. Guan, J. Li, L. Wu, Y. Zhang, J. Wu, X. Du, Achieving efficient and secure data acquisition for cloud-supported internet of things in smart grid. IEEE Internet Things J. **4**(6), 1934–1944 (2017)
16. D.L. Guidoni, G. Maia, F.S. Souza, L.A. Villas, A.A. Loureiro, Vehicular traffic management based on traffic engineering for vehicular ad hoc networks. IEEE Access **8**, 45167–45183 (2020)
17. M. Hao, H. Li, X. Luo, G. Xu, H. Yang, S. Liu, Efficient and privacy-enhanced federated learning for industrial artificial intelligence. IEEE Trans. Indust. Inform. **16**(10), 6532–6542 (2019)
18. J. Hayes, O. Ohrimenko, Contamination attacks and mitigation in multi-party machine learning. arXiv preprint arXiv:1901.02402 (2019)
19. N. Hussain, P. Rani, Comparative Studied Based on Attack Resilient and Efficient Protocol with Intrusion Detection System Based on Deep Neural Network for Vehicular System Security. Distributed Artificial Intelligence: A Modern Approach, 217 (2020)
20. R. Ito, M. Tsukada, H. Matsutani, An on-device federated learning approach for cooperative anomaly detection. arXiv preprint arXiv:2002.12301 (2020)
21. S. Jajodia, H.C. van van Tilborg (Eds.), *Encyclopedia of Cryptography and Security: A-K*. Springer (2011)
22. Y. Jiang, S. Wang, B.J. Ko, W.H. Lee, L. Tassiulas, Model pruning enables efficient federated learning on edge devices. arXiv preprint arXiv:1909.12326 (2019)
23. P. Kairouz, H.B. McMahan, B. Avent, A. Bellet, M. Bennis, A.N. Bhagoji, ..., S. Zhao, Advances and open problems in federated learning. arXiv preprint arXiv:1912.04977 (2019)

24. M. Kuderer, S. Gulati, W. Burgard, Learning driving styles for autonomous vehicles from demonstration, in *2015 IEEE International Conference on Robotics and Automation (ICRA)*. IEEE (2015, May), pp. 2641–2646
25. D. Li, J. Wang, Fedmd: Heterogenous federated learning via model distillation. arXiv preprint arXiv:1910.03581 (2019)
26. T. Li, M. Sanjabi, A. Beirami, V. Smith, Fair resource allocation in federated learning. arXiv preprint arXiv:1905.10497 (2019)
27. G.L. Li, J. Wu, J.H. Li, et al., Service popularity-based smart resources partitioning for fog computing-enabled industrial Internet of Things. IEEE Trans Ind Inform **14**(10), 4702–4711 (2018)
28. L.Z. Li, K. Ota, M.X. Dong, Deep learning for smart industry: Efficient manufacture inspection system with fog computing. IEEE Trans. Ind. Inform. **14**(10), 4665–4673 (2018)
29. L.Z. Li, K. Ota, M.X. Dong, DeepNFV: A light-weight framework for intelligent edge network functions virtualization. IEEE Netw., in press (2018)
30. T. Li, A.K. Sahu, A. Talwalkar, V. Smith, Federated learning: Challenges, methods, and future directions. IEEE Signal Process. Mag. **37**(3), 50–60 (2020)
31. W.Y.B. Lim, N.C. Luong, D.T. Hoang, Y. Jiao, Y.C. Liang, Q. Yang, et al., Federated learning in mobile edge networks: A comprehensive survey. IEEE Commun. Surveys Tutorials **22**(3), 2031–2063 (2020)
32. A.K. Malhi, S. Batra, H.S. Pannu, Security of vehicular ad-hoc networks: A comprehensive survey. Comput. Secur. **89**, 101664 (2020)
33. M. Jaggi, V. Smith, M. Takac, J. Terhorst, S. Krishnan, T. Hofmann, M.I. Jordan, Communication-efficient distributed dual coordinate ascent, in *Advances in Neural Information Processing Systems* (2014), pp. 3068–3076
34. V. Mothukuri, R.M. Parizi, S. Pouriyeh, Y. Huang, A. Dehghantanha, G. Srivastava, A survey on security and privacy of federated learning. Futur. Gener. Comput. Syst. **115**, 619–640 (2021)
35. F. Mo, H. Haddadi, Efficient and Private Federated Learning Using Tee, in *EuroSys* (2019)
36. H.B. McMahan, E. Moore, D. Ramage, B.A. Arcas, Federated learning of deep networks using model averaging. arXiv preprint arXiv:1602.05629 (2016)
37. B. McMahan, E. Moore, D. Ramage, S. Hampson, B.A. Arcas, Communication-efficient learning of deep networks from decentralized data, in *Artificial Intelligence and Statistics*. PMLR (2017, April), pp. 1273–1282
38. A. Nilsson, S. Smith, G. Ulm, E. Gustavsson, M. Jirstrand, A performance evaluation of federated learning algorithms, in *Proceedings of the Second Workshop on Distributed Infrastructures for Deep Learning* (2018, December), pp. 1–8
39. S. Niknam, H.S. Dhillon, J.H. Reed, Federated learning for wireless communications: Motivation, opportunities, and challenges. IEEE Commun. Mag. **58**(6), 46–51 (2020)
40. S. Orme, Addressing Issues with Defense-in-Depth, APTs, and IoT with Active Cyber Defense Cycle and Cyber Resilience (Doctoral dissertation, Utica College) (2019)
41. M. Obaidat, M. Khodjaeva, J. Holst, M.B. Zid, Security and privacy challenges in vehicular ad hoc networks, in *Connected Vehicles in the Internet of Things*, (Springer, Cham, 2020), pp. 223–251
42. S.R. Pokhrel, J. Choi, Federated learning with blockchain for autonomous vehicles: Analysis and design challenges. IEEE Trans. Commun. **68**(8), 4734–4746 (2020)
43. L.H. Pham, Foundations of Adaptive Cyber Defense against Advanced Persistent Threats (Doctoral dissertation, George Mason University) (2020)
44. S. Rizvi, J. Willet, D. Perino, S. Marasco, C. Condo, A threat to vehicular cyber security and the urgency for correction. Procedia Comput. Sci. **114**, 100–105 (2017)
45. J. Santa, A.F. Gómez-Skarmeta, M. Sánchez-Artigas, Architecture and evaluation of a unified V2V and V2I communication system based on cellular networks. Comput. Commun. **31**(12), 2850–2861 (2008)
46. Rani, Preeti, Naziya Hussain, Rais Abdul Hamid Khan, Yogesh Sharma, and Piyush Kumar Shukla. "Vehicular Intelligence System: Time-Based Vehicle Next Location Prediction in Software-Defined Internet of Vehicles (SDN-IOV) for the Smart Cities." In Intelligence of Things: AI-IoT Based Critical-Applications and Innovations, pp. 35-54. Springer, Cham, 2021

47. V. Smith, C.K. Chiang, M. Sanjabi, A. Talwalkar, Federated multi-task learning. arXiv preprint arXiv:1705.10467 (2017)
48. F. Tramèr, A. Kurakin, N. Papernot, I. Goodfellow, D. Boneh, P. McDaniel, Ensemble adversarial training: Attacks and defenses. arXiv preprint arXiv:1705.07204 (2017)
49. S. Truex, N. Baracaldo, A. Anwar, T. Steinke, H. Ludwig, R. Zhang, Y. Zhou, A hybrid approach to privacy-preserving federated learning, in *Proceedings of the 12th ACM Workshop on Artificial Intelligence and Security* (2019, November), pp. 1–11
50. X. Wang, Y. Han, C. Wang, Q. Zhao, X. Chen, M. Chen, In-edge ai: Intelligentizing mobile edge computing, caching and communication by federated learning. IEEE Netw. **33**(5), 156–165 (2019)
51. J. Wu, M. Dong, K. Ota, J. Li, Z. Guan, Big data analysis-based secure cluster management for optimized control plane in software-defined networks. IEEE Trans. Netw. Serv. Manag. **15**(1), 27–38 (2018)
52. Z. Xie, S. Zhu, Q. Li, W. Wang, You can promote, but you can't hide: large-scale abused app detection in mobile app stores, in *Proceedings of the 32nd Annual Conference on Computer Security Applications* (2016, December), pp. 374–385
53. Q. Yang, Y. Liu, Y. Cheng, Y. Kang, T. Chen, H. Yu, Federated learning. Synthesis Lectures Artificial Intell. Machine Learning **13**(3), 1–207 (2019)
54. Z. Yu, J. Hu, G. Min, Z. Zhao, W. Miao, M.S. Hossain, Mobility-aware proactive edge caching for connected vehicles using federated learning. IEEE Trans. Intell. Transp. Syst (2020)
55. Y. Zhao, J. Zhao, L. Jiang, R. Tan, D. Niyato, Mobile edge computing, blockchain and reputation-based crowdsourcing IoT federated learning: A secure, decentralized and privacy-preserving system. arXiv preprint arXiv:1906.10893 (2019)
56. Z. Zhao, C. Feng, H.H. Yang, X. Luo, Federated-learning-enabled intelligent fog radio access networks: Fundamental theory, key techniques, and future trends. IEEE Wirel. Commun. **27**(2), 22–28 (2020)
57. J. Zhang, J. Wang, Y. Zhao, B. Chen, An efficient federated learning scheme with differential privacy in mobile edge computing, in *International Conference on Machine Learning and Intelligent Communications*, (Springer, Cham, 2019), pp. 538–550
58. T. Zhu, S.Y. Philip, Applying differential privacy mechanism in artificial intelligence, in *2019 IEEE 39th International Conference on Distributed Computing Systems (ICDCS)*. IEEE (2019, July), pp. 1601–1609
59. Xu, Guowen, Hongwei Li, Sen Liu, Kan Yang, and Xiaodong Lin. "Verifynet: Secure and verifiable federated learning." IEEE Transactions on Information Forensics and Security 15 (2019): 911–926
60. Pan, Lei, Xi Zheng, H. X. Chen, T. Luan, Huzefa Bootwala, and Lynn Batten. "Cyber security attacks to modern vehicular systems." Journal of information security and applications 36 (2017): 90–100
61. Cretu, Gabriela F., Angelos Stavrou, Michael E. Locasto, Salvatore J. Stolfo, and Angelos D. Keromytis. "Casting out demons: Sanitizing training data for anomaly sensors." In 2008 IEEE Symposium on Security and Privacy (sp 2008), pp. 81-95. IEEE, 2008
62. Wei, Kang, Jun Li, Ming Ding, Chuan Ma, Howard H. Yang, Farhad Farokhi, Shi Jin, Tony QS Quek, and H. Vincent Poor. "Federated learning with differential privacy: Algorithms and performance analysis." IEEE Transactions on Information Forensics and Security 15 (2020): 3454–3469

Security Issues and Solutions for Healthcare Informatics

Bipin Kumar Rai

1 Introduction

EHR systems are extremely craved for the structured unification of all pertinent medical data of an individual and to exhibit lifelong medical record. Various confidentiality threats of healthcare data are crucial which may be either from within the institution or outside by some intruder. Each healthcare unit, hospital, and clinic have their own information system for maintaining the patient's data.

Therefore, standards for data exchange are required [4], and electronic health records and data need to be standardized, including semantic interoperability (standards for exchange of patient's data among EHR systems) [39, 41]. Several solutions like openEHR, Consolidated Health Informatics (CHI) initiative, Certification Commission for Healthcare Information Technology (CCHIT), Healthcare Information and Management Systems Society (HIMSS), American National Standards Institute (ANSI), and HL7 are used for maintaining health records [18]. Google Health and Microsoft's HealthVault provide facility to users to store and manage all health records in one central place. Anyone can import her/his health information from doctors, hospitals, etc. Most of these services don't provide full control to the patients.

Smart card healthcare systems developed in European countries are not strongly privacy preserving as anyone can access a patient's information from a health card without her/his consent.

Indivo is the first patient controlled web-based healthcare system which provides options to own a secure complete medical record, integrating EHRs of different health centers [4, 13, 14, 26, 27, 34].

B. K. Rai (✉)
Department of IT, ABES Institute of Technology, Ghaziabad, Uttar Pradesh, India

S. P. Yadav et al. (eds.), *Federated Learning for IoT Applications*, EAI/Springer Innovations in Communication and Computing,
https://doi.org/10.1007/978-3-030-85559-8_12

An efficient protocol and architecture for electronic health record (EHR) is required which is not standardized yet [3, 11, 12]. As EHR has succeeding advantages related to cost-saving reduction of adverse drug events, collection of patient's healthcare data for research, efficiency in medical treatment, etc.

Access control mechanisms and applications related to e-prescription systems and other consumer-related healthcare services require a secure mechanism [5, 6, 14, 15, 17]. As we know that huge size of patients' data and its management become difficult, hence big data technology and cloud computing are required [2, 47]. In the future, blockchain technology seems to be more appealing in the field of healthcare [42–45].

IoT technologies enable doctors/hospital authorities to be more watchful and establish close connection with the patients [57].

2 Blockchain Technology-Based Solution for Healthcare Informatics

Data sharing of the information related to patients is very crucial. Storage of health data are now moving toward cloud. To provide privacy and security, blockchain technology comes in the picture [21, 49, 51, 53–55].

Azaria et al. [58] proposed a framework in which health authorities can participate in the system as blockchain miners. This system gives secure access to his/her medical history with proper permission management.

It maintains decentralization of medical records, smart contract mechanism, and POW-based consensus algorithm.

Kaur et al. [58] have proposed an decentralization architecture in which files are fragmented and transaction stored on the various nodes. Whenever the user requests for a transaction, verification is done using cryptographic techniques. Henceforth, a new block is added on the existing blockchain. It has been shown in Fig. 1.

Dubovitskaya proposed a solution based on blockchain, which have databases for medical data storage, nodes, and interfaces for application programming [58]. The proposed architecture has been shown in Fig. 2.

Liang et al. proposed solution which is patient-centric system; they have used the blockchain technology, which is immutable, decentralized, and accountable. It has a mobile application that is synchronized with wearable devices. To ensure performance scalability, tree-based method and batching method have been used by the authors. The proposed architecture has been shown in Fig. 3.

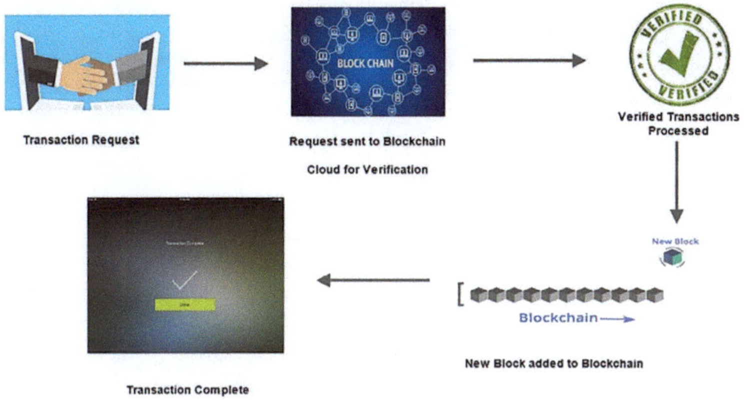

Fig. 1 Proposed architecture by Kaur et al. [58]

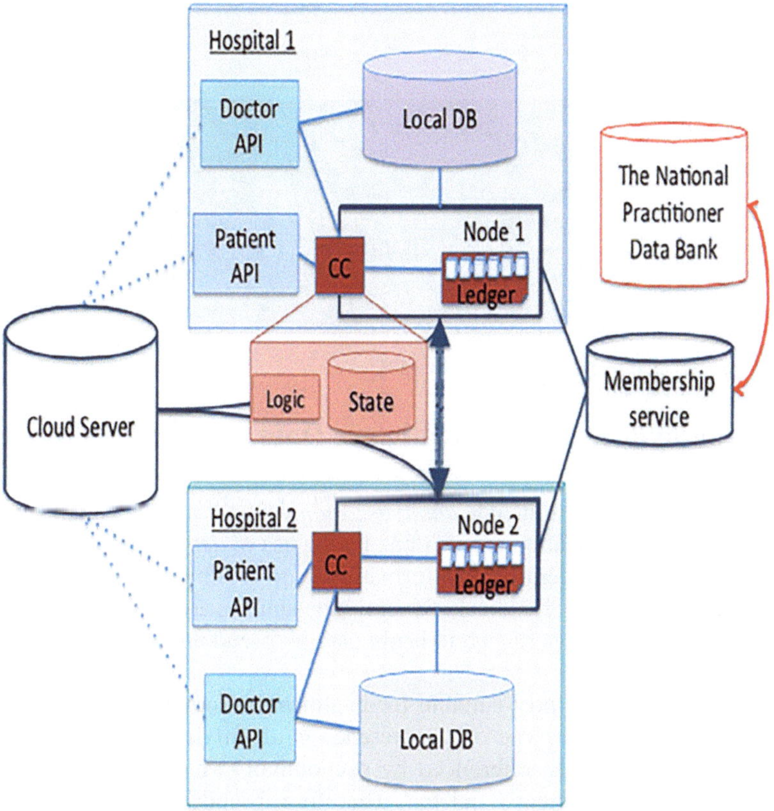

Fig. 2 Proposed architecture by Dubovitskaya et al. [58]

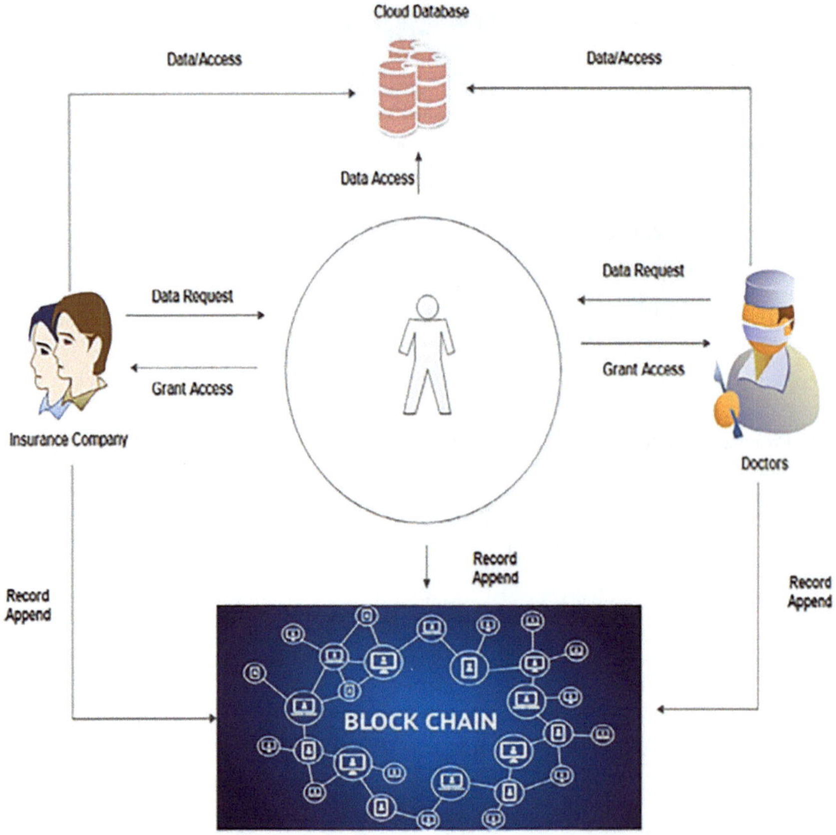

Fig. 3 Proposed architecture by Liang et al. [58]

3 Pseudonymization of Healthcare Data

Disassociation techniques include *anonymization* which is used nowadays for security purposes. Anonymization can be done by the use of *depersonalization* which is nothing but removal of identifying information from health records such that in any situation it cannot be reverted back. Several mechanisms are available for ensuring security and privacy issues related to healthcare by pseudonymiztion technique [7–10, 13–16].

Pseudonymization is most suitable for healthcare information system which is similar to anonymization. The only difference is identifying information is separated from health records, referenced by pseudonym (which is a unique random number), but not permanently deleted. Therefore, pseudonymization is patient's controlled reversible process under specified circumstances [22, 23].

Pseudonymization techniques have the potential to handle issues of identity management [40, 46]. It is also applied in several applications like healthcare information system. Several mechanisms based on pseudonymization are available for assuring privacy and security issues related to EHR [7, 9–11].

Pseudonymization-based solution for e-prescription system is very appealing. Confidentiality and integrity of data are needed in a safe system which can be ensured by pseudonymization technique [24, 25].

4 Access Control

The objective of access control is protecting any information system from unauthorized access at the same time making it available to authorized users [16, 20]. Electronic health record information systems need to develop a strong mechanism for protecting unauthorized access of data [36].

Obviously, the nature and constraints of the particular information system affect the access control methodology as well as architecture.

In literature, three standard access control mechanisms have been recognized. Each standard was designed to solve limitations found in its previous one [1].

4.1 DAC (Discretionary Access Control)

In this approach, each user accesses his/her data by his authorized identity [16]. Its limitations are:

(a) Difficult to manage.
(b) Might create new security issues [16].

4.2 MAC (Mandatory Access Control)

- In MAC mechanism, decisions are made by a CA (central authority), not by the owner of an object. Owner of data cannot change access rights [16].
- Medical personnel may assign access rights to healthcare entity, so some form of MAC policy should be involved.

  Its limitations are:

(a) Very difficult because of the huge number of user's medical records.
(b) Has limited scope because of unsuitability to patient control of their medical data.

4.3 RBAC (Role-Based Access Control)

Most existing access control mechanisms for healthcare sector are either RBAC-based mechanism or in some sense evolved from this mechanism. In RBAC, on the basis of the activities, role access to information are provided [28, 33, 35].

Its limitations are:

(a) Central access control module which can misused by administrator.
(b) Some evaluations of access requests are complex as it needs to consider contextual parameters.
(c) Insufficient for the requirements of healthcare systems.

These access control mechanisms like DAC, MAC, and RBAC are insufficient for EHRs, but combination of these can be used up to some extent [1, [19]].

4.4 Access Control Mechanism Suitable for the Healthcare System

4.4.1. **Privacy-aware role-based access control (P-RBAC)** is an extension of RBAC similar to XACML, which supports fine-grained policies [17].
4.4.2. **Personalized access control** uses combination of RBAC and DAC. In this approach, owner (patient) will decide who can access her/his health data.
4.4.3. **Context-related access control** for healthcare is RBAC with contextual parameters. It is not efficient to handle dynamic permission assignments [1, 32].
4.4.4. **Audit-based access control** – useful for urgent situation. It allows the medical officer to handle exceptional situations and justify later [1, 29].
4.4.5. **Behavior-based access control** is useful for handling the model compatible with several healthcare systems [31].
4.4.6. **Rule-based access control** approach is for resource-limited pervasive healthcare system.
4.4.7. **XACML-based access control –** Fig. 4 shows the sequence of steps.

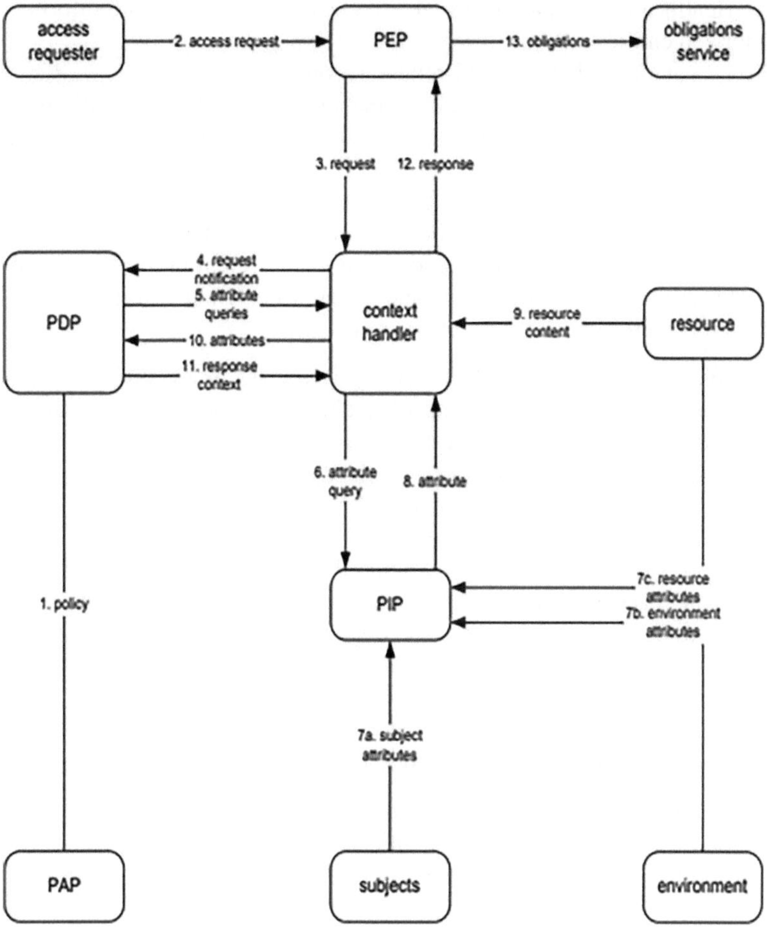

Fig. 4 Data flow of XACML-based access control [48]

5 E-Health Model in Serbia

In Serbia, The architecture of the healthcare system is a hybrid smart card-based solution [4]. The organization of the database and two models one online and second offline is briefly outlined here. In this system, RFID- and IC-based smart cards have been used for national level project.

It includes RFID antenna and microchip. Microchips can receive sufficient amounts of information. Chip cards are a kind of small computing machines which can perform calculations and exchange information to/from the system.

The RFID part of the smart card manages procedures for registration, authentication, as well as treatment. Basic components of this proposed system are:

1. HIS_Healthcare information system – it ensures access of health information by authorized entities only.
2. EHR – contains patient's health-related record.
3. EHR_applications – this application manages creation of patient's record, all history of activity like scheduling proposals, patient's visits to doctors, treatment done by doctor, and other services.
4. PHR – it is a combination of several online tools.
5. Card for Patient and Doctor – this electronic card is an alternative for the traditional health booklet used in hospitals, but it is different in form and functionality. This card is capable of storing and transmitting health information.
6. eRecipe – doctor issues an electronic recipe.

In Serbia, the architecture of electronic healthcare is a centralized system. A certificate authority manages issuing of health cards and renewal of certificates. Through the WAN network, all healthcare facilities or their healthcare information system would be connected to a central database. Patients can access their own health information from a central database by web portal.

The network used in this system works on client-server model.

The server is the central database. Standard protocols such as TCP/IP, IPSec, and SSL are being used for security. The database is implemented in SQL server. For security, new devices and systems are implemented to restrict access to certain information as per need. Different views can be created according to user rights as well as roles in the system.

All the healthcare data in the database can be used by the ministry for the purpose of monitoring and controlling drugs. The main tables are:

"Patient" table is the central part where all data of patients are stored.
"health_id_card" table – contains patient's data of all diseases.
"Therapy" – table for patient's therapy-related data and doctor's prescription. It is used to control unauthorized usage of drugs.
"Check" – where all doctor's appointments are stored. This table can check the effectiveness of doctors. Patients' ID card have debts-related data which can be checked as well.

An e-receipt is placed on his ID which can be read by a pharmacist using a card reader.

Only the prescribed medicine will be visible to save sensitive data. Whatever drug is purchased by the patient will be entered.

Any individual cannot purchase it without renewing doctor's prescription.

Doctor's prescription will be stored in database and in the ID of the patient.

6 Anonymous e-Prescriptions

From the security point of view, healthcare systems have several characteristics for consideration. The healthcare system includes a heterogeneous set of institutions and hence sometimes has conflicting goals [50, 52]. The whole patient's experience of medical care is private. Hence, providing confidentiality of medicine prescriptions is important [5].

The pharmacist collects the information often stored in databases owned by hospitals. There is no federal law for privacy of patient's health records maintained by pharmacies. Hence, privacy of patients is compromised. Some example of misuse of patient data has been felt. Medical prescription is defined as both entities – "a token signed by a doctor and a process."

Two reasons for this importance are:

1. The prescription quite reveals the patient's medical history.
2. Nowadays, insurance companies related to healthcare have the fastest rising budget [5].

The prescription model consist of patient, doctor, pharmacist-the entity filling prescription, insurer-provide health benefits for enrolled patients, privacy officer, judge-responsible for legal authority, certification authority of the medical board-responsible for granting power to doctors to issue prescriptions.

In this system, the first issue is access control; second issue is to handle the use of different resources which can be managed under a pseudonym.

The descriptions of confidential prescription protocol are –.

Patient enrollment: Firstly, every patient needs to enroll in a health insurance plan. When patient receives this smart card, then he can show his status of enrollment to the doctor.

Patient's verification: After enrollment in a health insurance plan, his verification process is carried out.

Doctor certification: A digital certificate is issued by the CA to the doctors. This digital certificate includes details of privileges like issuing prescription, etc. The doctor can join one or more healthcare provider organizations. After the doctor had done an agreement, he would be contacted by his/her provider organization. PO issues a pseudonym to the doctor.

Pre-approval: It works this way: the doctor connects to a digital clearinghouse to obtain pre-approval to issue a digital prescription. A form for prescription mentioning any policies will be filled out.

Filling: (a) The patient gives his smart card to the pharmacist. (b) After reading patient's pseudonym, Prescription is submitted one at a time to smart card of patient which signs them for pharmacist. The pharmacist forwards a digital token and collects an e-payment.

Anonymous profiling: To prevent fraud, the "PBM" profile patients and "Healthcare providers" profile doctors. "PBM" collect anonymous patient's record together using the same pseudonym.

Pseudonym revocation: Only the PO would be able to open pseudonyms. It is hoped that revocation procedure would be established.

It describes a relevant and practical system for electronic payment system to provide privacy services to doctors and patient. It is useful, but still refinement in security assumption is needed.

7 Smart Card-Enabled e-Prescriptions System [6]

Both patient and doctor have security concern with this e-prescription data as other parties are involved and some parties may use for their benefits like marketing, etc. It has a strong proxy signature scheme which satisfies security requirement such as strong unforgettability, verifiability, and identifiability.

This solution provides three types of delegation schemes: full delegation, partial delegation, and delegation by warrant. These schemes firstly provides delegation by using strong algorithm and then signing and verification procedure. Also analysis of these schemes are discussed that how can all above security parameter can be fulfilled.

As in this system the smart card plays a very important role, it becomes a target point for attacks, and hence its security is of primary concern. Its memory is organized into four sections.

1. Secret section – written only once; data contains card manufacturer's PIN and card holder's master key.
2. Sensitive section – only occasional update; contains issuer's PIN and card holder's PIN.
3. Working section – it can be erased and rewritten.
4. Public section – no requirement for protection; anyone can read.

Patients can easily get prescription without any security crisis. So this proposed system is quite practical.

8 Patient-Controlled Pseudonym-Based EHR (PcPbEHR)

This PcPbEHR system is patient-centered which combines the pseudonymization techniques and encryption to provide an efficient mechanism.

PcPbEHR system consists of four components as shown in Fig. 5 [26, 27, 56].

1. Two separate databases, PcPbEHR Health Records database and PcPbEHR Patient Profile database.
2. Pseudonymization module.
3. Access control module for patients and access control module for doctor/healthcare authority.
4. Data access policy and sharing management.

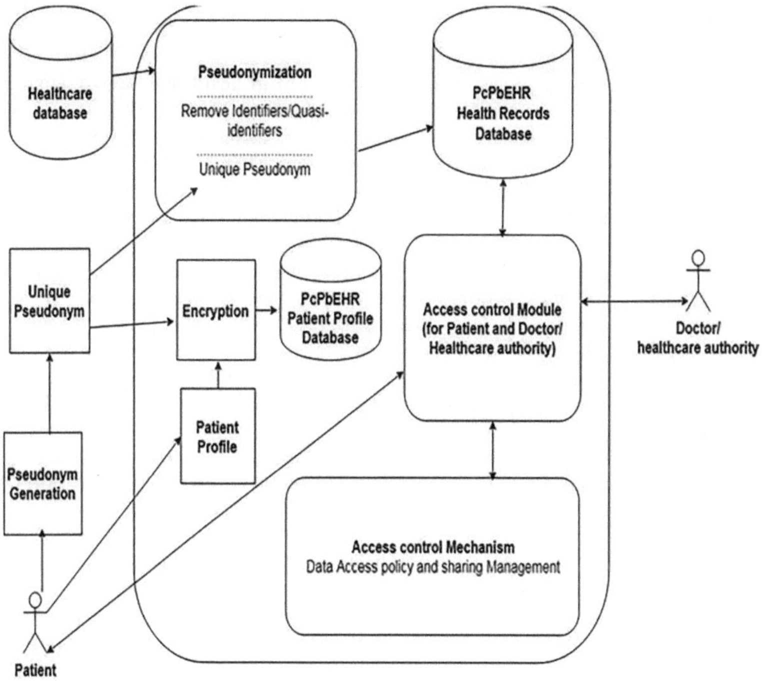

Fig. 5 Architecture of patient-controlled pseudonym-based mechanism of electronic health record (PcPbEHR) [26]

9 Conclusion

In this chapter, I have discussed several challenges of healthcare information system as well as some solutions for data management, access control, privacy preservation, and e-prescriptions. Most of the solutions advocate for the patient-centered system, a convenient, simple, privacy preserving, and secure mechanism.

The availability of patient's health information to any healthcare entity at any time with consent of patient is another very basic requirement. Pseudonymization-based solutions have great potential to provide a huge possibility for medical research on particular disease by using anonymous health data [30, 37, 38]. It can provide anonymous health data which will not have any identifiable information of the patient to the researchers. Hence, without compromising privacy, anonymous health data will be available for the researchers.

Blockchain technology-based mechanisms to deal with healthcare industry would allow users to transfer medical data in a protected way. This also provides users a way to anonymously share their medical data for the goal of medical research. Blockchain technology provides a secure way to share health data and also provides improved healthcare transactions.

References

1. W.A. Al-Hamdani, Cryptography based access control in healthcare web systems, in *2010 Information Security Curriculum Development Conference*, vol. 2010, (ACM)
2. R. Zhang, L. Liu, Security models and requirements for healthcare application clouds, in *2010 IEEE 3rd International Conference on cloud Computing*, (IEEE, 2010)
3. Huda, M.D. Nurul, N. Sonehara, S. Yamada, A privacy management architecture for patient-controlled personal health record system. J. Engineering Science and Technology, 4.2 154–170 (2009)
4. M. Vucetic, A. Uzelac, N. Gligoric, E-Health Transformation Model in Serbia: Design, Architecture and Developing, in *2011 International Conference on Cyber-Enabled Distributed Computing and Knowledge Discovery*, (IEEE, 2011)
5. G. Ateniese, B. de Medeiros, Anonymous e-prescriptions, in *Proceedings of the 2002 ACM Workshop on Privacy in the Electronic Society*, vol. 2002, (ACM)
6. Y. Yang et al., A smart-card-enabled privacy preserving E-prescription system. IEEE Trans. Inf. Technol. Biomed. **8**(1), 47–58 (2004)
7. B. Riedl et al., A secure architecture for the pseudonymization of medical data, in *The Second International Conference on Availability, Reliability and Security (ARES'07)*, (IEEE, 2007)
8. S. Poag, X. Deng, Information security and privacy concerns of online prescription systems. *Refereed research paper, okland University*
9. D. Slamanig, C. Stingl, Privacy aspects of e-health, in *2008 Third International Conference on Availability, Reliability and Security*, (IEEE, 2008)
10. B. Riedl, V. Grascher, Assuring integrity and confidentiality for pseudonymized health data, in *ECTI-CON2010: The 2010 ECTI International Conference on Electrical Engineering/ Electronics, Computer, Telecommunications and Information Technology*, (IEEE, 2010)
11. B. Alhaqbani, C. Fidge, Privacy-preserving electronic health record linkage using pseudonym identifiers, in *HealthCom 2008-10th International Conference on e-health Networking, Applications and Services*, (IEEE, 2008)
12. R. Addas, N. Zhang, Support access to distributed EPRs with three levels of identity privacy preservation, in *2011 Sixth International Conference on Availability, Reliability and Security*, (IEEE, 2011)
13. R. Peterson, "Encryption system for allowing immediate universal access to medical records while maintaining complete patient control over privacy". U.S. Patent Application No. 09/973,796
14. C. Thielscher et al., Patent: Data processing system for patient data. Int. Patent, WO **3.034294**, A2 (2005)
15. K. Pommerening, M. Reng, Secondary use of the EHR via pseudonymisation. Stud. Health Technol. Inform., 441–446 (2004)
16. L. Røstad, Access control in healthcare applications. Acc. Cont. Healthcare Informat. Sys. **37** (2008)
17. Q. Ni et al., Privacy-aware role-based access control. ACM Transactions on Information and System Security (TISSEC) **13.3**, 24 (2010)
18. E. Helms, L. Williams, Evaluating access control of open source electronic health record systems, in *Proceedings of the 3rd Workshop on Software Engineering in Health Care*, (ACM, 2011)
19. D.M. Eyers, J. Bacon, K. Moody, OASIS role-based access control for electronic health records. IEE Proceed. Software **153**(1), 16–23 (2006)
20. B. Alhaqbani, C. Fidge, Access control requirements for processing electronic health records, in *International Conference on Business Process Management*, (Springer, Berlin, Heidelberg, 2007)
21. S.P. Yadav, K.K. Agrawal, B.S. Bhati, et al., Blockchain-based cryptocurrency regulation: an overview. Comput. Econ. (2020). https://doi.org/10.1007/s10614-020-10050-0

22. B. Riedl et al., Pseudonymization for improving the privacy in e-health applications, in *Proceedings of the 41st Annual Hawaii International Conference on System Sciences (HICSS 2008)*, (IEEE), p. 2008
23. P. Vimalachandran et al., Ensuring data integrity in electronic health records: a quality health care implication, in *2016 International Conference on Orange Technologies (ICOT)*, (IEEE, 2016)
24. B.K. Rai, A.K. Srivastava, Security and Privacy Issues in Healthcare Information System. Inter. J. Emerg. Trends Technol. Comput. Sci. (IJETTCS) **3.6**, 2278–6856 (2014)
25. B.K. Rai, A.K. Srivastava, Pseudonymization Techniques for Providing Privacy and Security in EHR. Inter. J. Emerg. Trends Technol. Comput. Sci. (IJETTCS) **5.4**, 2278–6856 (2016)
26. B.K. Rai, A.K. Srivastava, Patient controlled Pseudonym-based mechanism suitable for privacy and security of Electronic Health Record. Inter. J. Res. Eng. IT Soc. Sci. (IJREISS) ISSN 2250 **7.2**, 588 (2017)
27. B.K. Rai, A.K. Srivastava, Prototype Implementation of Patient controlled Pseudonym-based mechanism for Electronic Health Record (PcPbEHR). Inter. J. Res. Eng. IT Soc. Sci. (IJREISS) ISSN 2250 **7.7**, 588 (2017)
28. K. Frikken, M. Atallah, J. Li, Attribute-based access control with hidden policies and hidden credentials. IEEE Trans. Comput. **55**(10), 1259–1270 (2006)
29. M.A.C. Dekker, S. Etalle, Audit-based access control for electronic health records. Electron. Notes Theoret. Comput. Sci. **168**, 221–236 (2007)
30. G.M. Bacelar-Silva et al., Comparing security and privacy issues of EHR: Portugal, the Netherlands and the United Kingdom, in *Proceedings of the 4th International Symposium on Applied Sciences in Biomedical and Communication Technologies*, (ACM, 2011)
31. M.H. Yarmand, K. Sartipi, D.G. Down, Behavior-based access control for distributed healthcare environment, in *2008 21st IEEE International Symposium on Computer-Based Medical Systems*, (IEEE, 2008)
32. D. Kulkarni, A. Tripathi, Context-aware role-based access control in pervasive computing systems, in *Proceedings of the 13th ACM symposium on Access control models and technologies*, (ACM, 2008)
33. P.T.B. Hue et al., An experimental evaluation for a new column–level access control mechanism for electronic health record systems. Inter. J. u-and e-Serv. Sci. Technol **4.3**, 73–86 (2011)
34. J. Benaloh et al., Patient controlled encryption: ensuring privacy of electronic medical records, in *Proceedings of the 2009 ACM workshop on cloud computing security*, (ACM, 2009)
35. Y. Zheng, Y. Chen, P.C.K. Hung, Privacy access control model with location constraints for XML services, in *2007 IEEE 23rd International Conference on Data Engineering Workshop*, (IEEE, 2007)
36. R. Gajanayake, R. Iannella, T.R. Sahama, Privacy Oriented Access Control for Electronic Health Records, in *Data Usage Management on the Web Workshop at the Worldwide Web Conference*, (ACM, 2012)
37. J. Heurix, M. Karlinger, T. Neubauer, Pseudonymization with metadata encryption for privacy-preserving searchable documents, in *2012 45th Hawaii International Conference on System Sciences*, (IEEE, 2012)
38. K. Garson, C. Adams, Security and privacy system architecture for an e-hospital environment, in *Proceedings of the 7th Symposium on Identity and Trust on the Internet*, (ACM, 2008)
39. S. Sachdeva, S. Bhalla, Semantic interoperability in standardized electronic health record databases. J. Data Informat Qualit (JDIQ) **3.1** (2012)
40. C.A. Kushida et al., Strategies for de-identification and anonymization of electronic health record data for use in multicenter research studies. Med. Care, S82–S101 (2012)
41. J. Jin et al., Access control model for sharing composite electronic health records, in *International Conference on Collaborative Computing: Networking, Applications and Worksharing*, (Springer, Berlin, Heidelberg, 2008)
42. A. Siyal et al., Applications of blockchain technology in medicine and healthcare: challenges and future perspectives. Cryptography **3.1**, 3 (2019)

43. C. Ananth, M. Karthikeyan, N. Mohananthini, A secured healthcare system using private blockchain technology. J. Eng. Technol. **6**, 42–54 (2018)
44. W.J. Gordon, C. Catalini, Blockchain technology for healthcare: Facilitating the transition to patient-driven interoperability. Comput. Struct. Biotechnol. J. **16**, 224–230 (2018)
45. D. Conceição, A. Flavio, et al., Electronic health records using blockchain technology. arXiv preprint arXiv, 1804.10078 (2018)
46. C. Quantin et al., Medical record search engines, using pseudonymised patient identity: an alternative to centralised medical records. Inter. J. Med. Informat. **80.2**, e6–e11 (2011)
47. H.A. Aziz, A. Guled, Cloud Computing and Healthcare Services (2016)
48. http://docs.oasis-open.org/xacml/3.0/xacml-3.0-core-spec-cs-01-en.pdf
49. A. Azaria, A. Ekblaw, T. Vieira, A. Lippman, MedRec: using blockchain for medical data access and permission management, in *Proceedings of the 2nd International Conference on Open and Big Data (OBD 16)*, (Vienna, Austria, 2016), pp. 22–24
50. J. Zhang, N. Xue, X. Huang, A secure system for pervasive social network based healthcare, in *IEEE Access*, vol. 4, (2016), pp. 9239–9250
51. X. Yue, H. Wang, D. Jin, M. Li, W. Jiang, Healthcare data gateways: Found healthcare intelligence on Blockchain with novel privacy risk control. J. Med. Syst. **40**, 218 (2016)
52. S.P. Yadav, S. Yadav, Image fusion using hybrid methods in multimodality medical images. Med. Biol. Eng. Comput. **58**, 669–687 (2020). https://doi.org/10.1007/s11517-020-02136-6
53. D. Ivan, Moving toward a blockchain-based method for the secure storage of patient records, in *ONC/NIST Use of Blockchain for Healthcare and Research Workshop*, (ONC/NIST, Gaithersburg, MD, USA, 2016)
54. S. Wang, J. Wang, X. Wang, T. Qiu, Y. Yuan, L. Ouyang, Y. Guo, F.Y. Wang, Blockchain-powered parallel healthcare systems based on the ACP approach, in *IEEE Transactions on Computational Social Systems*, vol. 5, (2018), pp. 942–950
55. S. Jiang, J. Cao, H. Wu, Y. Yang, M. Ma, J. He, Blochie: A blockchain-based platform for healthcare information exchange, in *Proceedings of 2018 IEEE International Conference on Smart Computing (SMARTCOMP)*, pp. 49–56
56. B.K. Rai, Pseudonymization based mechanism for security & privacy of healthcare, in *2020, ISBN (978-620-0-78791-0)*, (LAMBERT Academic Publishing, Germany)
57. S.P. Yadav, Emotion recognition model based on facial expressions. Multimed. Tools Appl. (2021). https://doi.org/10.1007/s11042-021-10962-5
58. A. Saha, R. Amin, S. Kunal, S. Vollala, S.K. Dwivedi, Review on "Blockchain technology based medical healthcare system with privacy issues". Secur. Priv. **2**(5), e83 (2019 Sep)

Federated Learning: Challenges, Methods, and Future Directions

Pushpa Singh, Murari Kumar Singh, Rajnesh Singh, and Narendra Singh

1 Introduction

Currently, in almost every enterprise and all walks of life, all new technologies such as blockchain, AI technology, and IoT illustrate their strengths [1, 18]. However, as we look back at the history of AI and blockchain, there have inevitably been some ups and downs in their development. Cellphones, wearable devices, and autonomous vehicles are only a few of the latest distributed networks that generate a wealth of data every day. Associated sensors on these devices, united with the point that they are always carried, indicate that they can have access to an extraordinary volume of data. However, the delicate nature of the data suggests that there are threats and duties in a centralized place to store it. Because of the growing processing power of these devices, as well as concerns about transmitting private information,

P. Singh (✉)
Department of Computer Science and Information Technology, KIET Group of Institutions, Ghaziabad, India
e-mail: pushpa.gla@gmail.com

M. K. Singh
Department of Computer Science and Engineering, Sharda University, School of Engineering and Technology, Greater Noida, India

R. Singh
Department of Information Technology, G.L Bajaj Institute of Technology and Management (GLBITM), Greater Noida, India

N. Singh
Department of Management Studies, GL Bajaj Institute of Management and Research, Greater Noida, India
e-mail: narendra.singh@glbimr.org

S. P. Yadav et al. (eds.), *Federated Learning for IoT Applications*, EAI/Springer Innovations in Communication and Computing,
https://doi.org/10.1007/978-3-030-85559-8_13

it is becoming increasingly desirable to store data locally and push network computing to the edge.

This situation has headed to a rising interest in federated learning (FL) [2].

FL is a method that does not involve sharing datasets with a central entity to train machine learning models. A model is collaboratively trained by various parties in FL, which retain their training dataset for themselves but engage in a shared FL process. The definition of parties could apply to entities as distinct as a company's data centers in various countries, clusters of computing in different clouds, mobile phones, vehicles, or other commercial entities. Google introduced FL technique that enables users to acquire the profits of connected models trained from this rich data collectively, without the requirement to store it centrally. The learning task is solved by a loose federation of contributing devices (cell phones, distributed devices) coordinated by a central device or servers [2].

Some famous toolkits like Tensorow Federated, TensorFlow Encrypted, Horovod, coMind, OpenMined, PaddleFL, and Clara Training framework, are used by Google [3]. Researchers introduced blockchain-based decentralized FL techniques because of data privacy criteria and the issues of centralization, accountability, and bias. Blockchain technology has gained extensive popularity for its transparency, distributivity, security, immutable ledgers, and autonomous distributed environment. Due to consensus mechanisms and the implicit features of non-repudiation, they allow trust among all participants [4]. BlockDeepNet incorporates blockchain and integrated IoT application learning algorithms, and each IoT system determines its local factors and trains its deep learning model [5]

This chapter attempts to clarify how FL works efficiently and discuss the types, architecture, applications, challenges, and future direction of FL. FL works in a heterogeneous environment and introduces several challenges and research direction [6].

The rest of the chapter is systematized as follows. Section 2 discusses the overview of FL. Section 3 introduces the foundation of machine learning workflow for FL. Challenges in FL have been given in Sect. 4. Future work directions are discussed in Sect. 5. Finally, Section 6 concludes the whole chapter.

2 Overview of Federated Learning

FL is a method that uses local data to download the current model and compute a modified model on the system itself, instead of going to a pool to update the device. These locally trained models are then sent back from the machines to the central server where they are aggregated and then returned to the devices with a single unified and enhanced global model. FL enables AI algorithms to gain experience from a wide range of data located at various locations.

Google is already using FL, which provides the Android keyboard (Gboard) with fantastic predictive input features, on-device search for Pixel phones, and other experiences. Google recently suggested the idea of FL. Google's main concept is to create dataset-based machine-learning models that are distributed across several devices in a secure manner. Recent developments have centered on reducing statistical challenges and enhancing FL security [2, 7]. FL techniques have been used by leading service providers to support privacy-sensitive systems where training data is decentralized at the edge, and they play an important role. FL improves traditional, fully centralized approaches, and the costs and risks associated with sensitive data handling, functioning best in bandwidth and power-restricted environments and providing a simple, efficient platform for personalization on scale. It also puts users in control of their data while enabling developers to create intelligent apps that leverage insights from that data. It also has the opportunity to serve as the catalyst for potential market changes [8].

2.1 Definition and Working of Federated Learning

The aim of FL is to produce a common model for multisite machine learning. In general, in FL, two processes exist: (i) model training and (ii) model inference. Information may be exchanged between individuals, but not data, in the process of model training. Any secured private part of the data at every site is not disclosed by the exchange. The trained model may exist at one party or shared between various parties. The model is functional to a new data instance at the time of inference. A new patient in a business-to-business setting, whose diagnosis is made in different hospitals, may be received by a federated system of medical imaging. In this scenario, in making a forecast, the parties cooperate. Finally, to share the benefit generated by the collaborative model, there should be an equal value-distribution system. In order to make the federation viable, process design should be performed in such a way.

FL is the machine learning practice where data holders work jointly to train an M_{FED} model without gathering any D_I: $1 \leq I \leq N$, allowing V_{Cent} and V_{FED} to represent the measure of the success of a centralized M_{Cent} model and a federated M_{FED} model, respectively. Let μ be a real number that is non-negative. We state that the M_{FED} federal model for learning has a performance loss if:

$$|V_{Cent} - V_{FED}| < \sigma \tag{1}$$

This eq. (1) expresses the following intuition. When we create an ML model on distributed data sources using stable FL, the output of this model in future knowledge is approximately the same as that of the model that is constructed in uniting all data sources. We are able to achieve a less than a standard model for the FL system since federated data holders do not disclose their data to a central server or any other database owner. This supplementary protection and privacy assurance can be much more beneficial than the **σ** benefit reduction in accuracy. A central coordinating machine may or may not be included in a FL system depending on the application. Figure 1 shows an example of a coordinator in a federated learning architecture.

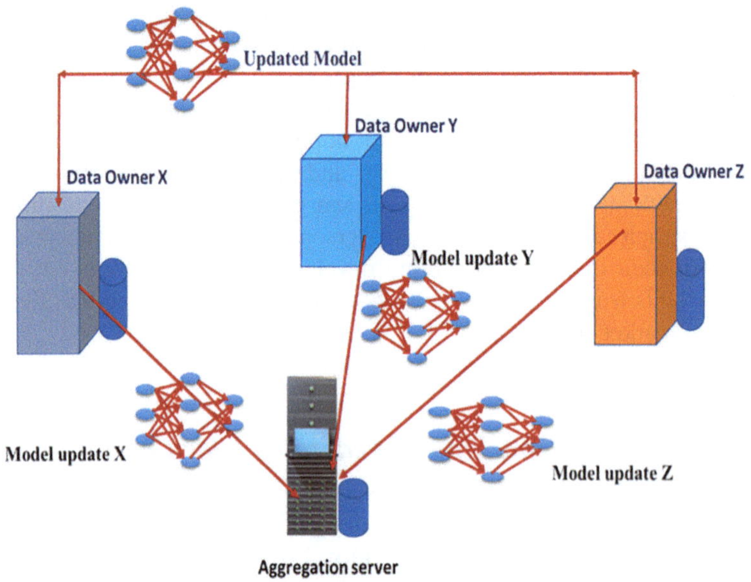

Fig. 1 Federated learning architecture: a client-server model

The coordinator is a central aggregation server that sends local data owners an initial model (X, Y, and Z). Using their respective datasets, local data owners X–Z train every model and send the model weight updates to the aggregation server. The aggregation is then combined with updates from data proprietors (e.g., with federated averages) [2] and return updates from the model to local data proprietors.

The above process is repeated until the pattern converges or until the maximum number of iterations is reached. The raw data of the local data owners under this architecture never leaves the local data holders. This method not only guarantees user secrecy and data safety but also protects the communication overhead required to send raw data. It is possible to encrypt the communication between the central aggregation server and local data holders to protect against leakage of information [7].

It is also possible to develop federal learning architecture in a peer-to-peer approach that does not rely onto coordinator or third-party mediator. As seen in Fig. 2, this also ensures security assurances in which the parties communicate directly without the assistance of any third party. Enhancement of security is one of the advantages of this architecture, but the downside is it required potentially more computation for encrypting and decrypting the messages.

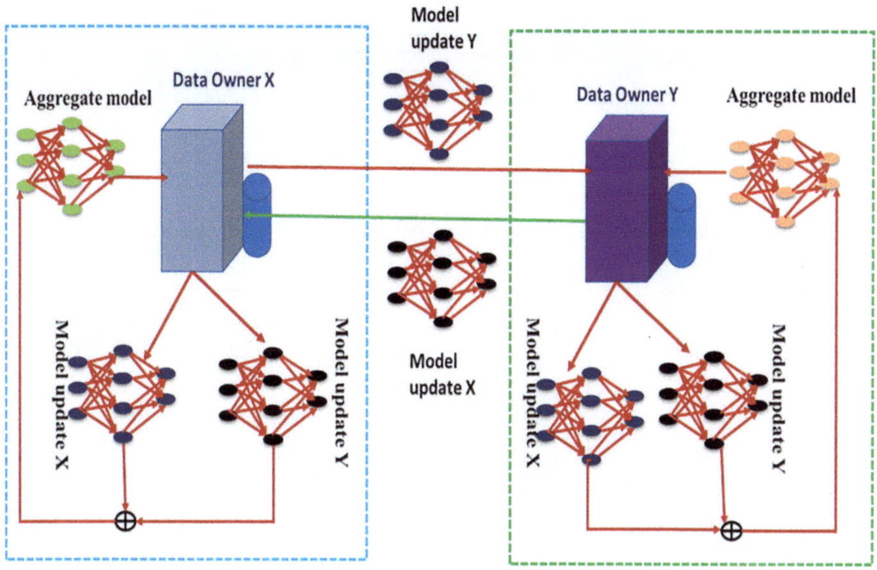

Fig. 2 A peer-to-peer federated learning architecture

2.2 *Types of Federated Learning*

Based on how data is divided between different function and sample space, as shown in Fig. 3, FL is graded into horizontal federated learning, vertical federated learning, and federated transfer learning. Figures 3 demonstrates the three categories of FL for a two-party scenario.

Horizontal Federated Learning (HFL): HFL is a federated sample analysis, in which the overlap of data collection held by various participants is used to train a model jointly. To predict labels for unlabeled samples of party X, a predictive model learned from feature representations of aligned samples longing to party X and party Y is used. Horizontal federated learning (HFL) refers to the situation where similar data features are shared by the participants in FL, i.e., the data features are compatible across the participants, but in data samples they differ. It is analogous to the situation in which the data inside a tabular view is horizontally partitioned. Therefore, we also refer to HFL (horizontal federated learning) as sample-partitioned FL, or FL partitioned by example.

Vertical Federated Learning (VFL): VFL refers to situations where participants share overlapping data samples in FL, i.e., data samples are matched with participants but vary in data features. It is similar to the situation where, within a tabular view, the data is vertically partitioned. We have also referred to VFL as federated feature-partitioned learning.

Federated Transfer Learning: This type of FL is applicable in cases where neither data samples nor features overlap. For example, if the two parties, X and Y,

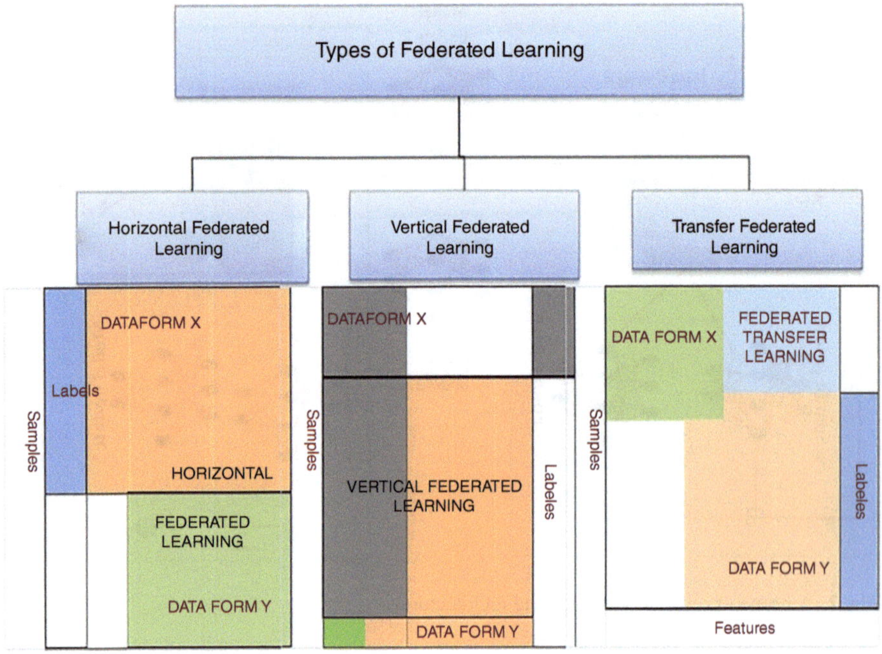

Fig. 3 Types of federated learning

are two banks operating in two separate regional markets, only a handful of users can share their data, but their data may be very similar to similar business models. That is, the two banks will work together to create ML models via HFL with restricted overlap in users but enormous overlaps in data features. [9, 10, 13].

2.3 Architecture of Horizontal Federated Learning

Two famous architectures for HFL systems are the client-server architecture and the peer-to-peer architecture.

2.3.1 Client-Server Architecture

A classic client-server architecture of an HFL is also known as master-worker architecture where K participants (or parties) with the same data structure collaboratively train a machine learning model with the support of an aggregation server or coordinator. This architecture assumes that the parties or participants are honest, but the aggregation server is curious because the aim is to prevent leakage of data or information from any participants [11]. Figure 4 demonstrates client-server architecture of a federated horizontal learning framework.

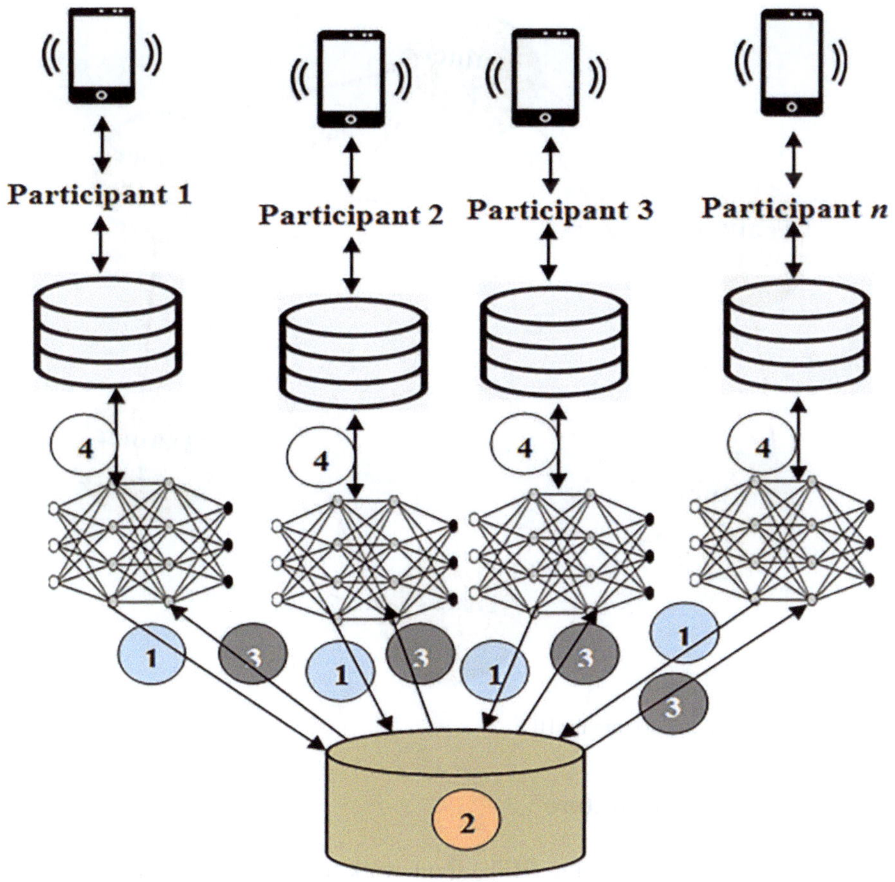

Fig. 4 Horizontal federated learning system: **A c**lient-server architecture

The preparation of such a federated horizontal training method consists generally of the following steps.

(i) Training gradients will be tested locally; gradients will be masked, encrypted, or secret shared technics will be used; and masked results will be submitted to the aggregation server.
(ii) A weighted average age is taken as a stable aggregation for the server.
(iii) The server returns the aggregated output to the participants.
(iv) The participants update their respective models with the decrypted gradients.
(v) In the above steps, the participants send gradients to the server, which in turn aggregates the received gradients also called gradient averaging [12].

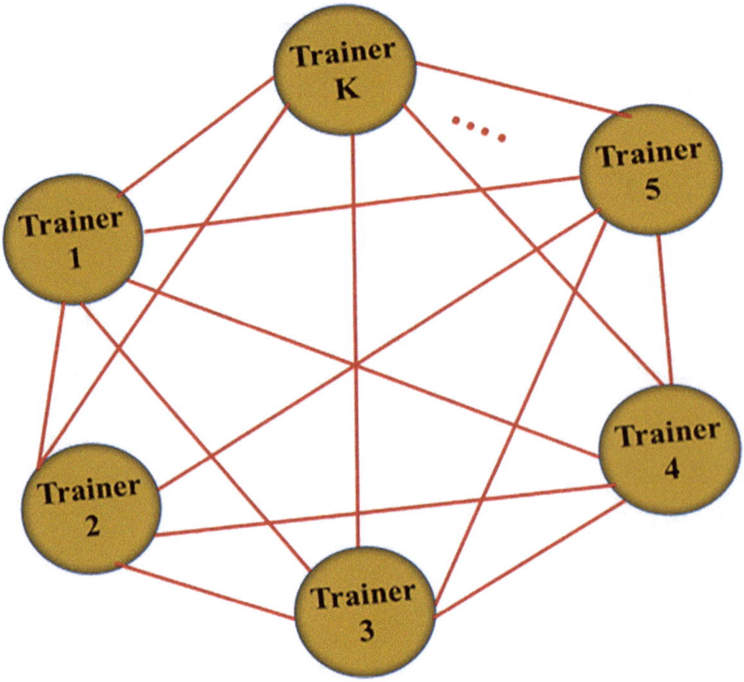

Fig. 5 Peer-to-peer architecture for HFL

2.3.2 Peer-Peer Architecture

A horizontal federated learning (HFL) framework may also make use of the peer-to-peer architecture, and under this, there is no central aggregation server or coordinator. Due to the lack of a central aggregation server, the trainers must agree on the order of sending and receiving the model weight in advance. In such cases, the K participants in an HFL system are often called distributed trainers. Each trainer is responsible for training the same ML model (e.g., a DNN model) using only its local data.

In peer-to-peer architecture, random transfer and cyclic transfer are two different approaches to send and receive the model weight.

2.4 Applications

FL eliminates the data sharing necessity for AI-based development. Recently several application areas are identified to have the potential of FL in various regions, as given below.

Healthcare The healthcare industry is the most important domain that will be significantly benefited from the rising of FL. Suppose all health institutions share their data such as disease symptoms, treatment plans, gene sequence, doctors' expertise, diagnostic report, etc. are integrated, and form a vast healthcare dataset. Then FL would be surely enhancing the performance of a trained model. FL demonstrated reliable performance when the distribution was highly skewed and imitating the actual scenario in which data distributions from various hospitals are diverse.

Transportation A real-world traffic dataset is used by Google Maps for the best route suggestion. FL enables to utilize the current information of all vehicles that crossed the same route in the last 1 hour and will be able to make a better decision in terms of safety and comfort of the passenger. FL also preserves privacy during the traffic flow prediction [13, 9].

Mobile Network Recently, mobile devices are fortified to integrate with cloud computing or edge computing due to advancements in deep learning. For any actual time-critical application, mobile edge computing has been essential to pass intelligence nearer to the edge. The concept of FL is used to train an AI-based model locally required by the central server [14, 24].

IoT and Blockchain FL has the ability to integrate with emerging technologies such as IoT and blockchain. Data can be collected from several home appliances using IoT devices. Model is trained with local data which establishes an initial model and sends it to blockchain. Blockchain provides a safe, transparent, and secure environment and replaces the centralized aggregator in the traditional FL system.

3 Life Cycle of Federated Learning Model

FL is a learning model seeking to address the various real-world problems by training algorithms collaboratively without exchanging the data itself. It is very significant to understand the process used to solve particular application. Well-established principles and life cycle models are essential to consider when developing a model of FL. This section offers life cycle of a model in FL as represented in Fig. 6.

(i). **Problem identification**: The first step is identification of the problem and its feasibility to be solved with FL.

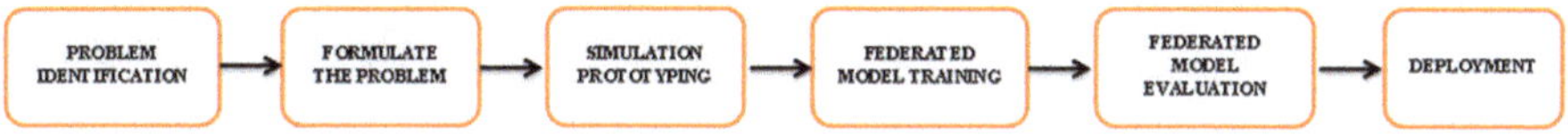

Fig. 6 Life cycle of a model in federated learning

(ii). **Formulate the problem with client equipment**: The next step is to formulate the problem and design the model to minimize computation times and maximize the effectiveness of training ML models. The model is designed with a different arrangement, storage capacity, user interface, etc. of client equipment as a proposed prototype model of the problem.

(iii). **Simulation prototyping**: Designed prototype model is simulated in a FL simulation using a proxy dataset.

(iv). **Training of model**: After the successful simulation of model, the next step is to train the model with real-world data. Several FL training-related jobs have begun to train diverse deviations of the model.

(v). **Federated model evaluation:** Selected model will go for the next step, i.e., model evaluation. These models are evaluated and trained in real environments for several days in order to select good candidates.

(vi). **Deployment:** After finalizing the best model, the next step is the implementation of the system on the best selected model. It goes through a standard model launch process, comprising manual quality assurance, live testing, and a staged rollout. The detailed launch method for a model is set by the proprietor of the application and is not dependent on the process of the model training.

4 Challenges in Federated Learning

FL is a new type of AI designed for the distributed and heterogeneous environment of their periphery for model training. FL is at its infancy and has little faith in the research community, as shown in Fig. 7, due to the following challenges and issues.

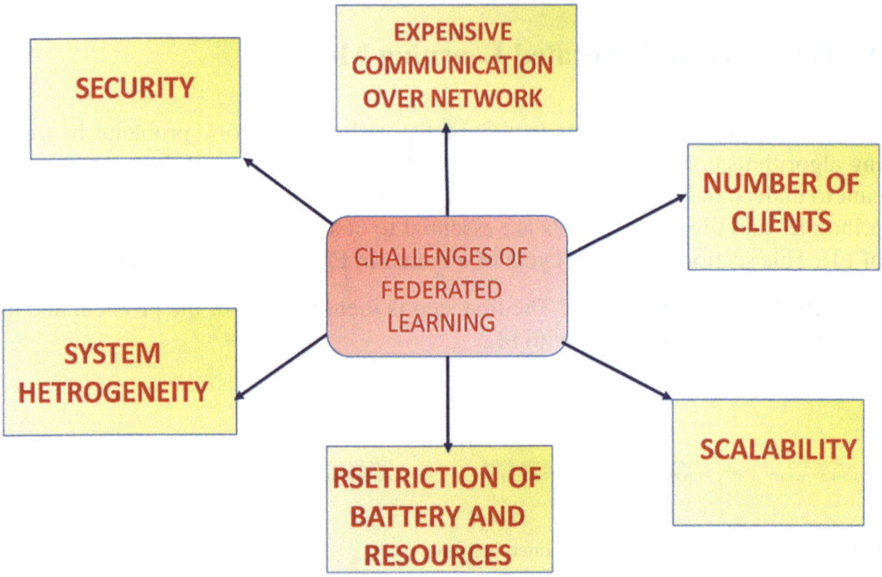

Fig. 7 Challenges of federated learning

4.1 Expensive Communication

FL refers to the creation of a unique, global statistical model from data stored on potentially millions of remote devices connected to the network through wired and wireless connections. Due to limited resource (bandwidth) and the increasing number of user devices and equipment, the communication network will be slower [15, 16] than local computation. Communication in such networks can be much more expensive than that in classical data center environments. The implementations of federated training approaches on traditional mobile networks incur expensive communication.

In order to deal with expensive communication, there is a need to develop an intelligent communication protocol that can efficiently send small training messages rather than the entire dataset over the network. Edge computing provides the potential to the FL to overcome communication issues.

4.2 Systems Heterogeneity

A present network is viewed as multiple level of heterogeneity in terms of hardware, network, devices, data storage, application, and battery level. Network heterogeneity means different types of network such as WLAN, WMAN, WWAN, and WPAN. The heterogeneity of the device includes various types such as mobile phones, smartphones, tablets, laptops, or other handheld devices that can communicate with other devices [17]. This overall heterogeneity becomes a challenge in FL. Furthermore, each system can use a heterogeneous data storage structure, and the independently and identically distributed (I.I.D.) assumptions can be violated, making the model more difficult to analyze. Moreover, data on the contributor devices are collected by the devices themselves, so there is a huge possibility of different data distributions among all the participants as each individual participant device collects the data based on its personal usage pattern and its local environment which might be different from other participants [19].

4.3 Number of Clients

In FL, the number of participants (client) is important to store and evaluate the collaborated learning models. These clients can often deny to participate in the training intentionally or unintentionally. Intentionally means the client is not interested in training and unintentionally is due to a poor network problem, no resource available, poor battery backup, etc. Managing these numbers of clients is not an easy task and hence becomes a big challenge [20]. It is required to ensure the presence of the participant in FL.

4.4 Scalability

Scalability is also one the common challenges in FL since a number of participants may increase a certain threshold. However, communication using parameter server can reduce to single round for participants and the server [21]. It also decreases the communication cost per client. However, communication through parameter server remains a challenge for communication-efficient distributed training because the upload and download to/from the server require efficient compression in order to reduce communication cost, time, and energy consumption.

4.5 Restrictions of Battery and Resource

Participants in FL are generally mobile devices. Mobile devices are usually facing the challenges due to limited battery power and resources. In deep learning, a single iteration is also very costly as it relies on rich battery power and memory (resource) requirement [22]. Due to limited size of memory, it is quite challenging to learn each training model. However, fog computing may provides a buffer area placed between the data processing location and the data storage equipment in the network to process real time training data [23].

4.6 Security

Security is also a major challenge in FL. FL security is related to participant (client) and communication network can violet the basic security requirement such as authentication, integrity, confidentiality, etc. of FL. Several network security threats such as malware, Trojan horse, viruses, spyware, worms, and phishing are critical attacks for FL. In fact, participating client can reveal confidential information to intruder, any third party, and fake central server.

FL must take initiative to protecting user data by sharing model updates instead of raw data. Secure multiparty computation and differential privacy are some techniques that can offer privacy and model efficiency at a low cost.

5 Future Directions

FL is a functioning and continuous zone of research. Albeit ongoing efforts started to address the difficulties talked about in this post, there are various basic open directions in FL that are however to be investigated. However, we briefly plan a combination of capable research bearings encompassing the recently examined difficulties (costly correspondence, frameworks heterogeneity, measurable heterogeneity, and security concerns) and present extra difficulties seeing issues, for example, benchmarking and productionizing in federated settings.

We are however planning to provide briefly a combination of research resources covering the difficulties recently studied, such as expensive correspondence, heterogeneity of frames, measureable heterogeneity, and security concerns.

Extreme Communication Schemes It stays not yet clear how much correspondence is fundamental in FL. In fact, it is notable that enhancement techniques for machine learning are able to endure an absence of exactness; this fault can truth be told assist through speculation. While communication plans have been investigated in anticipated server core settings for one-shot or partitioning and overcome, the conduct of such techniques is not certain of huge or measurable heterogeneous organizations. Also, one-shot/barely any shot heuristics [25–27] have as of late been proposed for the federated setting, however still can't seem to be hypothetically examined or assessed at level.

Communication Reduction and the Pareto Frontier We examined a few different ways to lessen announcement in federated training, for example, nearby refreshing and model compression. To make a reasonable framework for FL, it is essential to see how these procedures make with each other and to deliberately investigate the compromise among exactness and communication for each approach. In the context of a similar communication financial plan and preference for a variety of communication/accuracy profiles, the more important processes will be upgraded to the Pareto frontier to achieve accuracy that takes precedence over some other methodology. Comparative exhaustive studies for productive neural network inference have been carried out [28] and are essential at communication decrease procedures for FL in a significant manner.

Novel Models of Asynchrony Mass synchronous and asynchronous methodologies are the two most common communication schemes used in distributed enhancement. Nonetheless, in federated networks, every gadget is frequently undedicated to the job needing to be done, and most gadgets are not dynamic on some random cycle. Interestingly, in federated system, every device is regularly sacrilegious to the main job, and majority of the machines are not dynamic on some random emphasis. As a result, it's worth looking into the implications of this more realistic gadget-centric communication scheme, in which each machine will choose when to "awaken" and cooperate with the central worker in a timed manner.

Heterogeneity Diagnostics Current mechanism has enumerated statistical factual heterogeneity through measurements, for example, neighborhood divergence and earth mover's distance [29]. This inspires the accompanying inquiries: However, these measurements can't be handily determined over the federated organization prior to preparing happens. The importance of these measurements persuades the open inquiries that follow: (i) Will there be simple diagnostics available to quickly determine the degree of heterogeneity in federated networks? (ii) Can diagnostics similar to measure framework-related heterogeneity be developed? (iii) Can recent or current definitions of heterogeneity be used to enhance the construction of federated streamlining strategies? (iv) Are there straightforward diagnostics that can be utilized to measure frameworks and realistic heterogeneity prior to preparing? (v) How to know these diagnostics be misused to additionally advance the intermingling of federated streamlining techniques?

Granular Privacy Constraints Privacy is typically defined at either a local or global level in relation to all devices within the organization. However, in practice, it may be necessary to characterize security on a more granular level, as security requirements can differ diagonally devices or even across information focuses on a single device. For example, Li et al. [30] newly projected test explicit protection ensures, accordingly giving a more fragile type of security in return for more precise models. Creating strategies to deal with mixed security limitations is a motivating and continuous way of future work. Would we be capable of characterizing further granular notions of security and developing techniques to deal with mixed (device explicit or test explicit) protection limitations?

Beyond Supervised Learning It is essential that the strategies discussed so far are developed with the purpose of supervised learning, i.e., they expect the information in the federated system to be fully marked.

By and by, a large part of the information produced in sensible federated networks might be unlabeled or feebly named. Moreover, the current issue may not be to fit a model to information, yet rather to play out some exploratory information examination, decide total insights, or run a more intricate undertaking, for example, reinforcement learning. The handling of past supervised learning problems in federated networks probably will again entail comparative adaptability, heterogeneity, and protection challenges.

Productionizing Federated Learning There are many experimental issues that occurs when FL is at their production stage. These issues are related to the concept of drift, devices are represented different behavior at different time stamp and cold start problems [31]. These research challenges are future opportunities for the researchers since FL is still in its beginning stage. LEAF is a modular benchmarking framework for experimenting the concept of federated learning which consists open source federated datasets [32].

6 Conclusion

In this chapter we have introduced the basic idea, types, architecture, applications, challenges, and opportunities of federated learning. We have discussed horizontal and vertical types of federated learning which is based on client server and peer-to-peer architecture. Federated learning is recognized as one of the most demanding machines to learn how to preserve privacy in decentralized datasets, model aggregation methods, reputation-conscious models, privacy protection methods, IoT, and blockchain integration. At the end, we have defined various challenges which mainly occur due to heterogeneous and distributed environment. System and device heterogeneity, expensive communication, security, scalability, etc. are some emerging challenges and potential for future research effort. These challenges will involve interdisciplinary work from a comprehensive set of research societies. We hope that in the near future, federated learning will create a milestone where data and information could be united together and maintain privacy, fairly distributed, and available at each point of request.

References

1. P. Singh, N. Singh, Blockchain with IoT and AI: A review of agriculture and healthcare. Inter. J. Appl. Evolut. Computat. (IJAEC) **11**(4), 13–27 (2020)
2. B. McMahan, E. Moore, D. Ramage, S. Hampson, B.A. Arcas, Communication-efficient learning of deep networks from decentralized data, in *Artificial intelligence and statistics*, (PMLR, 2017), pp. 1273–1282
3. U. Rehman, M. Habib, K. Salah, E. Damiani, D. Svetinovic, Towards Blockchain-based reputation-aware federated learning, in *IEEE INFOCOM 2020 - IEEE conference on computer communications workshops*, vol. 2020, (INFOCOM WKSHPS, 2020), pp. 183–188
4. S.M.H. Bamakan, A. Motavali, A.B. Bondarti, A survey of blockchain consensus algorithms performance evaluation criteria. Expert Syst. Appl. **113385** (2020)
5. J. Konečný, H.B. McMahan, F.X. Yu, P. Richtárik, A.T. Suresh, D. Bacon, Federated learning: Strategies for improving communication efficiency. arXiv preprint arXiv:1610.05492 (2016)
6. T. Li, A.K. Sahu, A. Talwalkar, V. Smith, Federated learning: Challenges, methods, and future directions. IEEE Signal Process. Mag. **37**(3), 50–60 (2020)
7. H. Yu, S. Yang, S. Zhu, Parallel restarted SGD with faster convergence and less communication: Demystifying why model averaging works for deep learning, in *Proceedings of the AAAI Conference on Artificial Intelligence, Vol. 33, no. 01*, (2019), pp. 5693–5700
8. M. Ammad-Ud-Din, E. Ivannikova, S.A. Khan, W. Oyomno, Q. Fu, K.E. Tan, A. Flanagan, Federated collaborative filtering for privacy-preserving personalized recommendation system. arXiv preprint arXiv:1901.09888 (2019)
9. Y. Liu, Y. Kang, C. Xing, T. Chen, Q. Yang, A secure federated transfer learning framework. IEEE Intell. Syst. **35**(4), 70–82 (2020)
10. Q. Yang, Y. Liu, Y. Cheng, Y. Kang, T. Chen, H. Yu, Federated learning. Synth. Lect. Artific. Intellig. Mach. Learn. **13**(3), 1–207 (2019)
11. Y. Aono, T. Hayashi, L. Wang, S. Moriai, Privacy-preserving deep learning via additively homomorphic encryption. IEEE Transact. Informat. Forensic. Secur. **13**(5), 1333–1345 (2017)
12. H. Tang, S. Gan, C. Zhang, T. Zhang, J. Liu, Communication compression for decentralized training. arXiv preprint arXiv:1803.06443 (2018)

13. Y. Liu, J.Q. James, J. Kang, D. Niyato, S. Zhang, Privacy-preserving traffic flow prediction: A federated learning approach. IEEE Internet Things J. **7**(8), 7751–7763 (2020)
14. W.Y.B. Lim, N.C. Luong, D.T. Hoang, Y. Jiao, Y.C. Liang, Q. Yang, et al., Federated learning in mobile edge networks: A comprehensive survey. IEEE Communications Surveys & Tutorials **22**(3), 2031–2063 (2020a)
15. P. Singh, R. Agrawal, A customer centric best connected channel model for heterogeneous and IoT networks. J. Organizat. End User Computing (JOEUC) **30**(4), 32–50 (2018)
16. R. Singh, N. Singh, A. Dinker, Performance analysis of TCP variants using AODV and DSDV routing protocols in MANETs. Recent Advanc. Comput. Sci. Communicat. **14**, 448 (2021)
17. P. Singh, R. Agrawal, A game-theoretic approach to maximize payoff and customer retention for differentiated services in a heterogeneous network environment. Int. J. Wirel. Mob. Comput. **16**(2), 146–159 (2019)
18. R. Singh, N. Singh, Performance assessment of DSDV and AODV routing protocols in Mobile Adhoc networks with focus on node density and routing overhead, in *2020 International Conference on Emerging Smart Computing and Informatics (ESCI)*, vol. 2020, (Pune, India, 2020), pp. 298–303
19. Y. Zhao, M. Li, L. Lai, N. Suda, D. Civin, V. Chandra, Federated learning with non-iid data. arXiv preprint arXiv:1806.00582 (2018)
20. T. Nishio, R. Yonetani, Client selection for federated learning with heterogeneous resources in mobile edge, in *ICC 2019-2019 IEEE International Conference on Communications (ICC)*, (IEEE, 2019), pp. 1–7
21. M. Asad, A. Moustafa, T. Ito, M. Aslam, Evaluating the communication efficiency in federated learning algorithms. arXiv preprint arXiv:2004.02738 (2020)
22. F. Sattler, S. Wiedemann, K.R. Müller, W. Samek, Robust and communication-efficient federated learning from non-iid data. IEEE Transact. Neural Networks Learn. Sys. **31**(9), 3400–3413 (2019)
23. Singh, P., & Agrawal, R. (2021). An Overloading State Computation and Load Sharing Mechanism in Fog Computing. Journal of Information Technology Research (JITR), 14(4), 94-106.
24. W.Y.B. Lim, N.C. Luong, D.T. Hoang, Y. Jiao, Y.C. Liang, Q. Yang, et al., Federated learning in mobile edge networks: A comprehensive survey. IEEE Communicat. Sur. Tutor. **22**(3), 2031–2063 (2020b)
25. M. Yurochkin, M. Agarwal, S. Ghosh, K. Greenewald, N. Hoang, Y. Khazaeni, Bayesian non-parametric federated learning of neural networks, in *International Conference on Machine Learning*, (PMLR, 2019), pp. 7252–7261
26. N. Guha, V. Smith, Model aggregation via good-enough model spaces. arXiv preprint arXiv:1805.07782 (2018)
27. N. Guha, A. Talwalkar, V. Smith, One-shot federated learning. arXiv preprint arXiv:1902.11175 (2019)
28. T. Bolukbasi, J. Wang, O. Dekel, V. Saligrama, Adaptive neural networks for efficient inference, in *International Conference on Machine Learning*, (2017)
29. S. Vaswani, F. Bach, M. Schmidt, Fast and faster convergence of sgd for over-parameterized models and an accelerated perceptron, in *The 22nd international conference on artificial intelligence and statistics*, (PMLR, 2019), pp. 1195–1204
30. J. Li, M. Khodak, S. Caldas, A. Talwalkar, *Differentially Private Gradient-Based Meta-Learning* (Technical Report, 2019)
31. Li, T., Sahu, A. K., Talwalkar, A., & Smith, V. (2020). Federated learning: Challenges, methods, and future directions. IEEE Signal Processing Magazine, 37(3), 50-60.
32. S. Caldas, S.M.K. Duddu, P. Wu, T. Li, J. Konečný, H.B. McMahan, A. Talwalkar, Leaf: A benchmark for federated settings. arXiv preprint arXiv:1812.01097 (2018)

Quantum Federated Learning for Wireless Communications

R. M. Pujahari and Akshit Tanwar

1 Introduction

Quantum federated learning (distributed learning) is nothing but a technique based on machine learning that uses quantum mechanics to solve problems. It uses several decentralized edge devices or servers to train an algorithm without sharing data samples. This method differs from conventional centralized machine learning methods, which require all local datasets to be submitted to a single server, as well as traditional approaches which are not centralized, on the basis of assumption that local samples of data are uniformly distributed, as well as more traditional centralized approaches, which require all local datasets to be submitted to a single server.

Quantum federated learning enables multiple actors to construct a scalable machine learning model without sharing data, addressing critical issues such as data privacy, security, data access rights, and access to heterogeneous data. Protection, telecommunications, IoT, and pharmaceutics are only a few of the industries where it's used.

Federated learning entails an algorithm of machine learning in the form of training, as an example neural networks which are deep, using many datasets which are in the local nodes stored, without specifically sharing samples of data. The idea is that parameters are exchanged and local models are trained on local data samples. The local models are trained on samples on local data and periodically sharing variables (e.g., the biases related to deep neural network and weights) among these local nodes to create and design model related to all nodes which is globally accepted.

R. M. Pujahari (✉)
ABES Institute of Technology, Ghaziabad, India

A. Tanwar
Macquarie University, Macquarie Park, NSW, USA
e-mail: tanwar.akshit@students.mq.edu.au

S. P. Yadav et al. (eds.), *Federated Learning for IoT Applications*, EAI/Springer Innovations in Communication and Computing,
https://doi.org/10.1007/978-3-030-85559-8_14

The key difference among the distributed learning and federated learning is about the local datasets and their characteristics [1] as opposed to learning in distributed format, which was designed to streamline computing resources, while federated learning was designed to train on heterogeneous datasets.

In the case of learning in distributed mode, often the objective is to train a model which is unique and single across several servers, an assumption which individually ascertains that the datasets are local and uniformly distributed (i.i.d.) and of comparable size. Since datasets are generally heterogeneous and can span several orders of magnitude in size, not a single hypothesis is appropriate for the purpose of learning in federated mode.

Furthermore, federated learning clients may be unreliable due to their reliance related to communication media which are not efficient enough (such as Wi-Fi) and the systems which are battery-powered (as an example smart devices like mobile phones and IOT-operated devices) compared to distributed learning, where nodes are connected to each other through faster networks and are generally nothing but data centers with powerful computing capabilities.

There is an increasing interest in using data-driven machine learning (ML)-based solutions to complement traditional model-based design approaches in the wireless communications environment. Traditional machine learning approaches assume that data and processing heads are housed in a central entity, but this isn't always the case. Due to the inaccessibility of private data and the high transmission overhead needed to transmit raw data to central ML processors, this is not always feasible in wireless communications applications.

As a result, decentralized machine learning methods that store data where it is generated are much more appealing. Due to its privacy-preserving nature, federated learning is especially important for many wireless applications, particularly in the context of fifth generation (5G) networks.

This chapter introduces the general concept of quantum federated learning, discusses some possible implementations in 5G networks, and identifies key technical issues and open problems for future quantum federated learning research in the wireless communications context.

Future wireless communications is expected to utilize machine learning for various tasks, such as security issue [30]. Federated learning aims to tackle these issues by distributing the burden of learning into edge computation units [31]. That is, instead of uploading all of the training data to the central processing unit, the data is kept in the edge devices, thus enhancing data privacy. Moreover, QNN gains research attentions amid performance gain offered by quantum computations [32, 33]. Motivated by these factors, this study inspects the usage of QFL for wireless communications, which utilizes multiple QNNs. QFL can be used for general optimization in wireless (such as optimizing power allocation in non-orthogonal multiple access [34]).

1.1 Centralized Quantum Learning in Federated Mode

The quantum federated learning which works in centralized environment, a server operates centrally orchestrates the various steps of the algorithms and coordinates all of the nodes which take part in the process of learning. In this case the server is in charge of node selection at the start of the training phase, as well as the aggregation of updates obtained related to various models. A single entity where all of the nodes which are chosen must submit updates, the server instead working as a device roadblock [2].

1.2 Decentralized Quantum Federated Learning

A global model is obtained by organizing the nodes in a decentralized quantum federated learning environment. Instead of synchronization of the server which is central, this configuration checks failures at single point, and the model alterations are basically shared among the nodes which are interjoined. Nonetheless, the learning process's output can be influenced by the network topology [2, 3].

1.3 Heterogeneous Federated Learning

A range of application domains, such as mobile phones and IoT applications, include a large number of heterogeneous clients. The majority strategies of current federated learning presume the models at the local level having the architecture at the same global model. Hetero-FL is a new federated learning platform with which clients at heterogeneous level are addressed with varying computational and capabilities with communication abilities [4]. Hetero-FL allows you to train local models at heterogeneous level which varies dynamically with the computation complexities while still generating an inference model which is unique and global [4].

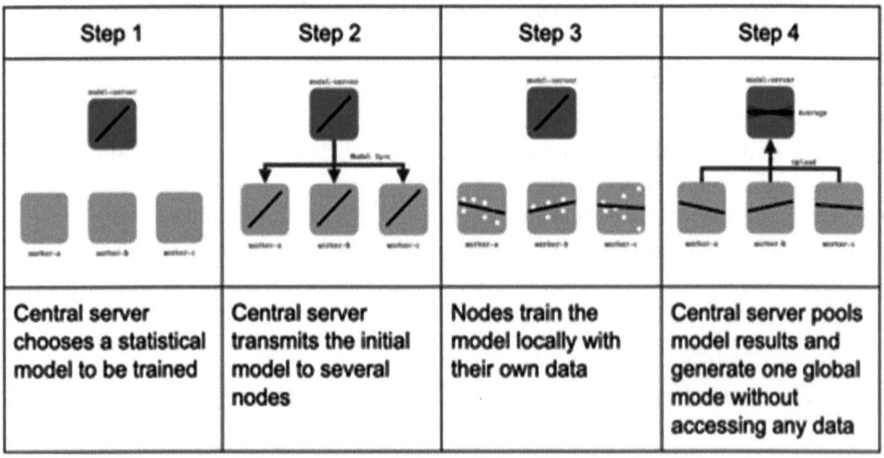

In a central orchestrator setup, federated learning is a general operation

2 Main Features of Quantum Federated Learning

2.1 Learning Iterative Nature

In case of quantum federated learning for the credebility of a good performance of the task, the whole process of iterative learning is segregated into a sequence of communication between server and client. So also to summarize the output, each node is looped, and the resultant sum of outcome is finalized for the output [2].

In the technique as described, to sum up, a server at the central level is used, while local nodes carry out training at the local level in response to the commands by the central server. Without the use of central servers, other approaches, such as a peer-to-peer strategy either with gossip [5] or using the methodologies of consensus, produce the same results [6]. The learning method can be summarized as follows, assuming one reference to the learning process in a federated round [7].

Initialization: Based on the feedback by servers, a model based on machine learning (e.g., linear regression, neural network, boosting) is selected and operated on local nodes. The nodes are then switched on, and the calculation tasks are listed to be allocated by the central server.

Selection of client: At the start of training on local data, a subset of local nodes is chosen. The current statistical model is acquired by the selected nodes, while the rest of the world waits when the adjacent quantum federated round shall start.

Configuration: A subset of nodes is directed to train the local data model in a predetermined order by the central server (e.g., descent in small batches for updates).

Reporting: The local models are sent toward the server by every selected node for summing up. The server at the center, which summarizes the models received, sends model updates and returns them back to the nodes. Errors caused by disconnected nodes or incomplete model updates are also addressed. Returning to the client selection process, the next quantum federated round has begun.

Termination: In case a previously defined finalization condition is achieved, the server at the center summarizes the updates to conclude the model at the global level (e.g., an optimum number of iterations is achieved, or the accuracy at model level exceeds a benchmark).

In the method as discussed previously which assumes that the model changes are coordinated and organised. New techniques for dealing with asynchronicity during the training phase, or having training with models which are dynamically varying, have been implemented in recent quantum federated learning developments [4]. Asynchronous approaches, in contrast to synchronous approaches, which exchange local models after the neural network strata have been computed, when the computations for a particular strata are derivable the model updating is done taking note of the situation. The already mentioned methods are correlated with each other and termed as segregated and split learning [8, 9], and all of them can be used for inference and preparation both in centralized and decentralized quantum federated learning environments [2].

2.2 Non-iid Data

The principle of separate and distributed equally as samples related to nodes at local level does not obey in most federated learning models. The probability distribution of training examples and the unbalanced nature of local data samples (i.e., labels and features) conserved at local nodes may have a major impact on the performance of the training process. Again exploring the outcomes of non-iid findings, the review as follows considers the important categories as described by Peter Kiarouz and colleagues in 2019 [2]. The data at non-iid level is represented using an assessment of the combined probability between labels and features for every node. This feature distinguishes the nodes and recognises the contributions from the available distribution of the nodes at local level.

Covariate shift: Local nodes will keep track of statistical distributions that aren't the same as those of other nodes. Processing of datasets in natural language is done, for example, people write the identical letters or digits with altogether using different slants or stroke widths [2].

Prior probability shift: Labels with statistical distributions that differ from those of other nodes may be stored in local nodes. This can occur if datasets are partitioned by area and/or demographics. Animal image datasets, for example, vary a lot from one country to the next [2].

Concept shift (*different features, same label*): While certain labels are shared by local nodes, few of them have different node characteristics. Images depicting a specific object, for example, will differ depending on the weather conditions under which has been taken [2].

Concept shift (*different labels, same features*): While nodes at the local level can have similar characteristics, some of them have different labels corresponding to various nodes. Even if the same text is observed, sentiment analysis in natural language processing can yield different sentiments [2].

Unbalancedness: The amount of available data at the level of local nodes can change in a wide range [2, 4]. Non-iid data illustrators are used to calculate the topology shifts of dynamic network [10] triggered by dataset changes or local node failures or ineligibility during the quantum federated learning process. Nodes involved during the training phase of learning in the global model might not be concluded due to inadequate calculating capacities. As a result, the training and analysis of data sample statistics differ.

3 Hyper-Parameter Algorithm

3.1 Topology Related to Network

As described earlier about the centralized model, the way statistical local outputs are gathered and communication between the nodes would be different. As a consequence, there are many federated learning methods to choose from, including no central administered server and communication related to stochastic method [11]. The basic significant difference is to execute and highlight the least circulated networks. As central server is not there, in this case, there is requirement of summarizing the local model and transmitting queries to local nodes. Every local node is made responsible for sending its outputs to a random group of nodes, who then aggregate the results at local level. This reduces the number of purchases, which can cut down on computing costs and training time [12].

3.2 Quantum Federated Learning Parameters

If the node network's topology has been determined, federated learning factor mechanism (as opposed to own machine learning model hyper-factors) can be managed for optimization of the learning as follows:

1. Number of quantum federated learning rounds.
2. Total number of nodes used in the process.
3. Fraction of nodes used at each iteration for each node.
4. Local batch size used at each learning iteration.
5. Factors related to model related can also be modified.
6. Iterations number for local training before pooling.
7. Local learning rate.

These parameters must be optimized based on the machine learning application's constraints (e.g., available computing power, available memory, bandwidth). Similarly the stochastic gradient descent is monitored by selecting every iteration of small fraction of nodes which as a result reduces computational cost and also prevents over-fitting.

4 Quantum Federated Learning Variations

Let us add some notations [13] to explain federated strategies:

The number of data samples accessible to the client during training.
The weight vector of the model on the client during the federated round.
For weights and batches, there is a loss feature.
The cumulative local epochs quantity.

4.1 Quantum Federated Stochastic Gradient Descent (FedSGD)

The majority of training related to deep learning focuses over stochastic gradient descent types, in which gradients are computed on a random subset of the total dataset and then used to create a single gradient descent stage.

Quantum federated stochastic gradient descent [14] is a direct conversion of this algorithm to a quantum federated environment, with a random fraction of nodes and all data on each node. The server uses the gradients to perform a gradient descent stage, averaging them equally for the quantity of training samples for every node.

4.2 Quantum Federated Averaging

FedAvg (quantum federated averaging) is a FedSGD generalization that permits local nodes to function as several group updates over locally collected data and not to prefer gradients but to prefer updated weights. The logic behind this theory is that in case each and every local node starts from the equal initialization, aggregating the gradients is very firmly analogous to averaging the weights themselves in FedSGD. In addition, aggregating the molded weights from the referred initialization would not always diminish the aggregate model's efficiency [13].

5 Limitations on Technical Ground

During the process of learning, quantum federated learning necessitates regular contact between nodes. It needs not only a good number of local computing of memory and power but also bandwidth which is very optimum to be able to part with the parameters of model related to machine learning. The technology, on the other hand, avoids the dissipation of data, which can require a lot of resources before centralized machine learning can begin. The machines utilized in federated learning, on the other hand, are generally communication restricted, such as smartphones or devices based on IoT or those that are typically linked to networks by Wi-Fi. As a result, mechanisms related to federated learning may not be obsolete in their current format, even though models are usually less expensive to transmit than raw data [2].

There are a variety of statistical problems with quantum federated learning:

Local dataset heterogeneity: each node can be biased in contrast to the general population, and dataset sizes can vary remarkably [4].
Temporal heterogeneity: the distribution of each local dataset can change over time.
The datasets related to every node must be compatible; this is a requirement.
Curations will be required on a regular basis for each node's dataset.
Attackers may be able to insert backdoors into the global model if training data is hidden [15].
Due to a lack of global training data, it is more difficult to detect unnecessary differences in the training, such as age, gender, and sexual orientation.
Node failures affecting the global model cause model changes to be lost in part or completely.

6 Quantum Federated Learning Properties

6.1 Design Privacy

The federated machine-related key benefits for learning methods is that they protect data protection or confidentiality. Local data at local level are neither exchanged nor uploaded nor concentrated with the rest of the planet. It is more difficult to break into the database since it is segmented into local bits.

In quantum federated related learning, only parameters related to machine learning are shared.

Additionally, parameters of such type can be decoded before being exchanged between learning cycles to improve anonymity, and encryption schemes which are homomorphic can be used to perform computations straight with the decoded data having to decrypt it first. Despite these precautions, these parameters can still expose information about the underlying data samples, for example, by conducting several queries on various datasets. As a consequence, node querying is an important focus, which is solved with privacy and restore aggregation [16]. Quantum federated learning's privacy concerns are also caused by running calculations, which prevents advanced deep learning models from being used. Static batch normalization (sBN) was designed to optimize privacy-constrained deep neural networks. sBN does not log running estimates during the training process, instead normalizing batch data. After the model converges, only the statistics of secret representations from local data are measured. Since local models do not need to upload running estimates during preparation, this approach is appropriate for the FL system. After optimization, local models only upload their statistics once, greatly reducing the chance of data leakage.

6.2 *Personalization*

The created model offers inputs on the basis of patterns of nodes at the global level. If a participating node in a multi-task learning system needs to learn from global patterns while also adjusting particular situation result, the learning approach related to federated module might be changed to create duel models at a time. After the learning process is done, clustering strategies may be used to group nodes that have certain similarities. This helps the nodes to generalize the models they've learned based on their local data [17].

For deep neural related networks, it is necessary to part certain layers related to nodes while keeping others on each local node. In most cases, the first layers, which perform general recognition of pattern, are trained and shared across all sets of data. The final resultant layers will be on every local node and will be learned only on that node's set of data [18].

Earlier personalization approaches often add extra computation and communication overhead that isn't always needed. A "Masking Trick" technique was built in FL to dramatically reduce computing and communication costs [4]. When compared to classical quantum federated learning, the "Making Trick" allows local clients to contribute adaptively to the training of global models much more flexibly and efficiently.

6.3 *Legal Issues of Quantum Federated Learning*

Data security and data traceability are becoming increasingly important in Western legal structures. The implementation of a data lessening theory, and stated in the European GDPR, was recommended in the White House 2012 Report [19] [20]. Transferring data from one country to another is illegal in certain cases (e.g., genomic related data), but foreign organized functions are also needed for advancement in scientific way. In reference to such instances, quantum federated learning offers suggestions for training a global model while staying stable.

Quantum federated learning provides strategies for training a universal model while remaining stable in such situations.

7 Current Research Topics

Quantum federated learning became a major research topic after the initial imprint on federated aggregating in the initiation of telecommunication modalities in 2015 [1] and 2016 [21]. The effective study is the decrease of communication overload in the course of learning process of federated mechanism.

In 2017 and 2018, literature and established facts stressed on the design and development of allocation of amenities and related strategies, especially to minimize the communication [13] needs [21] among various nodes with gossip algorithms [22], so also the characterization of the robustness to various individual privacy attacks [23]. Various research focuses on quantization and sparsification techniques [21], in which machine learning models are sparsified and/or compressed before being parted with different nodes to minimize training bandwidth.

Recent developments in science are beginning to embrace real-word propagating channels, rather than ideal channels [24], which were assumed in previous implementations. Federated learning, which is used to train local models that are stratified with different computation jargons into a unique efficient global outcome model, is another active research area [4]. Assisted learning is a modern learning framework that aims to enhance each agent's efficiency while keeping models, individual data, and learning goals hidden [25]. As opposed to quantum federated learning, which also involves a controller at the central level for the learning and optimization, induced learning tends to supply protocols for the agents to maximize and learn. Supported learning seeks to include standard procedures for agents to refine and learn among themselves in the absence of a model at global level.

8 Cases in Use

Quantum federated related learning is used when individual players essentially conceptualized models on general sets of data than their own but can't risk the data to be shared with anyone (e.g., for the reasons might be economic, legal, or strategic). Servers at the local level and the interconnection between them and a least amount of computing power for nodes are still required by the technology [2].

8.1 Transportation: Self-Driving Cars

Self-driving cars use a variety of machine learning technologies to get around, including computer vision to detect obstacles and machine learning to adjust their speed in response for the environment (e.g., road unevenness). A traditional cloud approach could pose a security risk due to the large number of self-driven cars that could be produced, and the requirement is to react quickly to real-world situations. Quantum federated learning may be a way to reduce data transfer volume while also speeding up learning processes [26, 27].

8.2 Smart Manufacturing: Industry 4.0

Machine learning techniques are widely used in Industry 4.0 to increase the efficiency and efficacy of manufacturing processes while ensuring a high degree of protection. Nonetheless, for industries and manufacturing firms, the protection of sensitive data is important. Since quantum federated learning algorithms don't disclose any information, they can be used to solve these problems [29].

8.3 Medicine: Digital Health

Instead of sharing data through quantum learning which is federated and aims to work out the issue of privacy and data security concerns for patients. The current method of combining data from various sources comes at the expense of serious privacy and data security concerns for patients. A key technology for resolving this issue is the capability to inculcate models related to machine learning on several medical colleges without movement of data. The authors of "The Future of Digital Health with Federated Learning" [29] explored the way how federated learning could solve the future of digital health in *Nature Digital Medicine* in September 2020.

8.4 Wireless Communications

The operation of two-tier QNN circuits in QFL is presented in Fig. 1. Consider a wireless network where users are grouped into N_G groups. Each group, consisting of $U_{m,1}$ and $U_{m,2}$ in particular, is served by an access point 1.

Each access point E_m, $\forall m \in \{1, \ldots, N_G\}$, receives channel information of $U_{m,1}$ and $U_{m,2}$ ($\tilde{h}_{m,1}$ and $\tilde{h}_{m,2}$, respectively) from pilot signals. Subsequently, each E_m, $\forall m \in \{1, \ldots, N_G\}$, predicts the solution for Gm using a particular shallow QNN, Q_m^{local}. The local prediction of G_m, which is a set of optimization solutions in G_m, is denoted

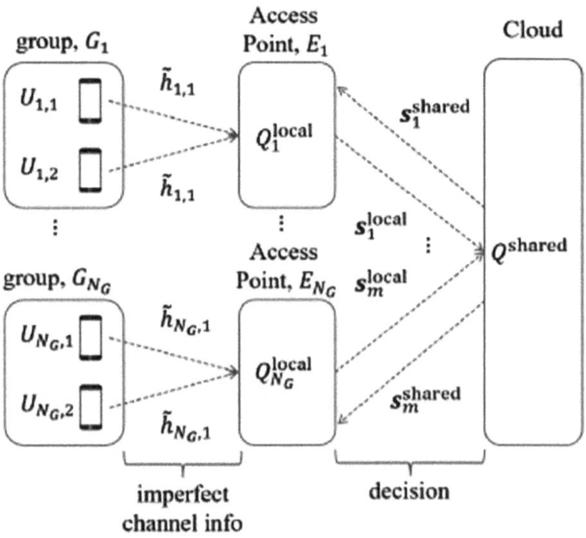

Fig. 1 QFL for wireless optimization, which consists of two tier operations: access point QNN (Q_m^{local} , m ∈ {1, . . ., N_G}) and cloud QNN (Q^{shared})

as $S_m^{\text{local}} = \{\lambda_{m,1}, \lambda_{m,2}\}$, where $\lambda_{m,k}$ is the optimization value for $U_{m,k}$. This study refers to "solution" as a general term for wireless optimization 2. Subsequently, the cloud processing unit takes s local m, ∀m ∈ {1, . . ., N_G}, as an input for shared QNN (i.e., Q_{shared}). The output of Q^{shared} is used to optimize E_m, ∀m ∈ {1, . . ., N_G}.

A. *Local QNN*

Figure 2 presents the circuit design for the local QNN. This study combines the multi-layer QNN design of [8] with activation function of [4] for local QNN. The local QNN, which is computed by E_m to predict S_m^{local} based on $h^{\sim}_m$, is defined as.

$$Q_m^{\text{local}} = \underbrace{\text{CCX}\left(ancilla\middle|q_1,q_2\right)}_{\text{"activation function"}} \underbrace{\text{X}^{\otimes N_G}}_{\text{"neuron connection"}} \text{H}^{\otimes N_G} U^{\text{encode}} U^{\text{local}} \text{H}^{\otimes N_G}, \tag{1}$$

where U^{encoding} and U^{local} are encoding [7] and unitary operation of shallow QNN.

Definition 1 (Shallow QNN) The unitary operations of shallow QNN are defined as.

$$U^{\text{local}} \triangleq \underbrace{\left(\text{R}_z\left(\theta_{m,k}^{\text{local}}\right)\right)^{\otimes N_q^{\text{local}}}}_{\text{"weight"}} \text{CZ}\left(q_2\middle|q_1\right), \tag{2}$$

where CCX(*ancilla*|*q1, q2*) indicates X-axis rotation (applied on an *ancilla* qubit, controlled by *q1, q2)* and CZ(*q2*|*q1*) indicates Z-axis rotation (applied on q_2, controlled by *q1*).

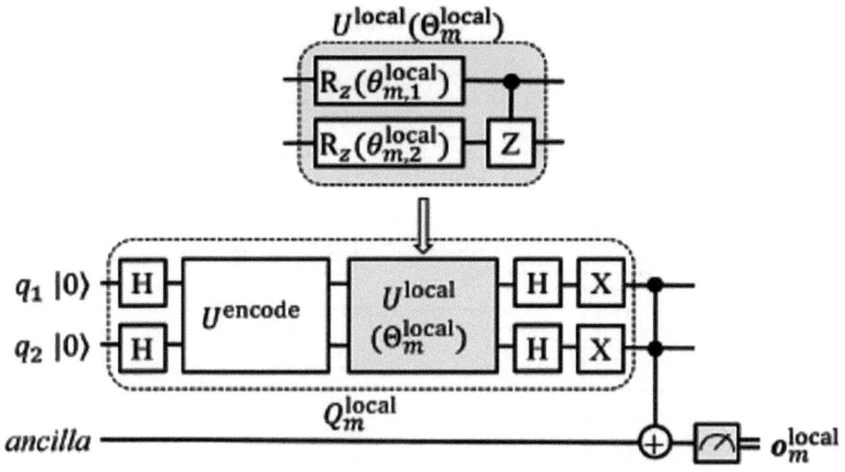

Fig. 2 Unitary operation for local QNN. The multi-layer QNN design of [8] is combined with activation function in [4]

B. *Shared QNN* Fig. 3 presents the circuit design for the shared QNN. The design of shared QNN in here is based on the multi-layer QNN design of [8] and activation function of [4]. The shared QNN, which is computed in cloud to predict $\mathbf{s}^{\text{shared}}$, is given as.

$$Q^{\text{shared}} = \left(\bigotimes_{i=((N_G-1)N_l)+1}^{N_G N_l} \text{XH} \right) U^{\text{shared}} U^{\text{encode}} \text{H}^{\otimes N_G N_l}, \tag{3}$$

where N_l indicates the number of layers. Moreover, U^{shared} denotes the operation of deep QNN, which is defined as follows.

Definition 2 (Deep QNN) The unitary operations of deep QNN is defined as.

$$U^{\text{shared}} \triangleq \underbrace{\left(\prod_{l=1}^{N_l} \text{CX}\left(q_{N_G l} \middle| q_{N_G(l-1)}\right) \right)}_{\text{"inter layer connection"}} \underbrace{\left(\prod_{l=1}^{N_{\text{neuron}}} \text{CZ}\left(q_{N_G l} \middle| q_{(N_G l)-1}\right) \right)}_{\text{"inter neuron connection"}} \underbrace{\left(\bigotimes_{l=1}^{N_l} \bigotimes_{m=1}^{N_G} \text{R}_z\left(\theta_{l,m}^{\text{local}}\right) \right)}_{\text{"weight"}}, \tag{4}$$

where $\text{CX}(q_{NGl}\ |q_{NG(l-1)})$ indicates X-axis rotation (applied on q_{NGl}, controlled by $q_{NG(l-1)}$).

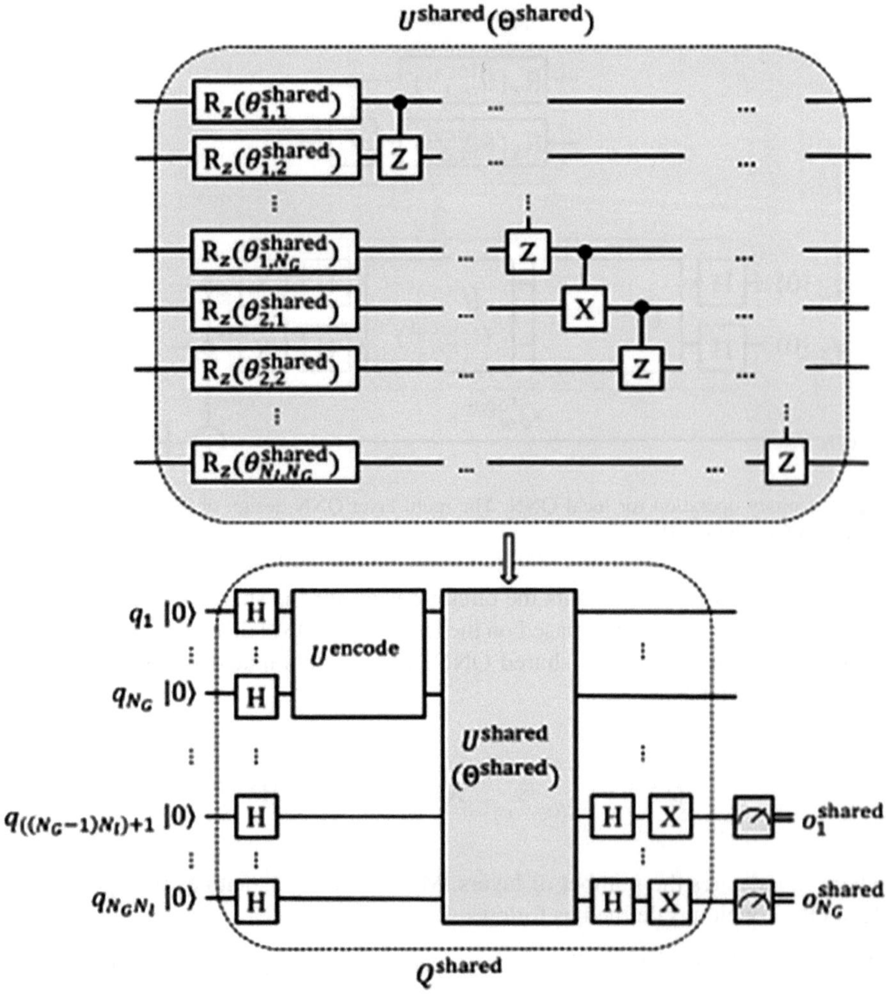

Fig. 3 Unitary operation for shared QNN. Similar to local QNN, the multi-layer QNN design in [8] and activation function in [4] are utilized

References

1. J. Konečný, B. McMahan, D. Ramage, Federated optimization: distributed optimization beyond the datacenter (2015) arXiv:1511.03575[cs.LG]
2. P, Kairouz, M.H. Brendan, B. Avent, A. Bellet, M. Bennis, N.B. Arjun, K. Bonawitz, Z. Charles, G. Cormode, R. Cummings, R.G.L. D'Oliveira, E.R. Salim, D. Evans, J. Gardner, Z. Garrett, A. Gascón, B. Ghazi, P.B. Gibbons, M. Gruteser, Z. Harchaoui, C. He, L. He, Z. Huo, B. Hutchinson, J. Hsu, M. Jaggi, T. Javidi, G. Joshi, M. Khodak, et al. Advances and open problems in federated learning (2019) arXiv:1912.04977 [cs.LG]

3. S.R. Pokhrel, J. Choi, Federated Learning with Blockchain for autonomous vehicles: Analysis and design challenges. IEEE Trans. Commun. **68**(8), 4734–4746 (2020). https://doi.org/10.1109/TCOMM.2020.2990686.S2CID219006840
4. E. Diao, J. Ding, V. Tarokh, HeteroFL: computation and communication efficient federated learning for heterogeneous clients (2020) arXiv:2010.01264 [cs.LG]
5. Decentralized collaborative learning of personalized models over networks. Paul Vanhaesebrouck, Aurélien Bellet, Marc Tommasi (2017)
6. S. Savazzi, M. Nicoli, V. Rampa, Federated learning with cooperating devices: a consensus approach for massive IoT networks. IEEE Internet Things J. **7**(5), 4641–4654 (2020). arXiv:1912.13163. https://doi.org/10.1109/JIOT.2020.2964162. S2CID 209515403
7. Towards federated learning at scale: system design, Keith Bonawitz Hubert Eichner and et al., (2019)
8. O. Gupta, R. Raskar, Distributed learning of deep neural network over multiple agents (2018) arXiv:1810.06060 [cs.LG]
9. P. Vepakomma, O. Gupta, T. Swedish, R. Raskar, Split learning for health: Distributed deep learning without sharing raw patient data (2018) arXiv:1812.00564 [cs.LG]
10. H. Eichner, T. Koren, H.B. McMahan, N. Srebro, K. Talwar, Semi-Cyclic Stochastic Gradient Descent (2019) arXiv:1904.10120 [cs.LG]
11. Collaborative deep learning in fixed topology networks, Zhanhong Jiang, Aditya Balu, Chinmay Hegde, Soumik Sarkar (2017)
12. GossipGraD: scalable deep learning using gossip communication based asynchronous gradient descent, Jeff Daily, Abhinav Vishnu, Charles Siegel, Thomas Warfel, Vinay Amatya (2018)
13. Communication-efficient learning of deep networks from decentralized data, H. Brendan McMahan and et al. (2017)
14. Privacy preserving deep learning, R. Shokri and V. Shmatikov (2015)
15. How to backdoor federated learning, Eugene Bagdasaryan (2018)
16. Practical secure aggregation for privacy preserving machine learning, Keith Bonawitz (2018)
17. F. Sattler, K.-R. Müller, W. Samek, Clustered Federated Learning: Model-Agnostic Distributed Multi-Task Optimization under Privacy Constraints (2019) arXiv:1910.01991 [cs.LG]
18. M.G. Arivazhagan, V. Aggarwal, A.K. Singh, S. Choudhary, Federated learning with personalization layers (2 December 2019) arXiv:1912.00818[cs.LG]
19. Anonymous, consumer data privacy in a networked world: a framework for protecting privacy and promoting innovation in the global digital economy. J. Privacy Confidentiality (2013) https://doi.org/10.29012/jpc.v4i2.623
20. Recital 39 of the Regulation (EU) 2016/679 (General Data Protection Regulation)
21. K. Jakub, H. Brendan McMahan, D. Ramage, P. Richtárik, Federated optimization: distributed machine learning for on-device intelligence (2016)
22. J. Konečný, H.B. McMahan, F.X. Yu, P. Richtárik, A.T. Suresh, D. Bacon, Federated Learning: Strategies for Improving Communication Efficiency (2017) arXiv:1610.05492 [cs.LG]
23. Gossip training for deep learning, Michael Blot and al., (2017)
24. Differentially Private federated learning: a client level perspective Robin C. Geyer and et al., (2018)
25. M.M. Amiri, D. Gunduz, Federated learning over wireless fading channels (2020) arXiv:1907.09769 [cs.IT]
26. X. Xian, X. Wang, J. Ding, R. Ghanadan, Assisted learning: a framework for multi-organization learning. Advances in Neural Information Processing Systems **33**, arXiv:2004.00566 (2020)
27. S.R. Pokhrel, Federated learning meets blockchain at 6G edge: a drone-assisted networking for disaster response 49–54 (2020) https://doi.org/10.1145/3414045.3415949. S2CID 222179104
28. A.M. Elbir, S. Coleri, Federated learning for vehicular networks (2020) arXiv:2006.01412 [eess.SP]
29. R. Cioffi, M. Travaglioni, G. Piscitelli, A. Petrillo, D. Felice, Fabio, Artificial intelligence and machine learning applications in smart production: progress, trends, and directions. Sustainability. 12(2), 492 (2019). https://doi.org/10.3390/su12020492

30. N. Rieke, J. Hancox, W. Li, F. Milletarì, H.R. Roth, S. Albarqouni, S. Bakas, M.N. Galtier, B.A. Landman, K. Maier-Hein, S. Ourselin, M. Sheller, R.M. Summers, A. Trask, D. Xu, M. Baust, M.J. Cardoso, The future of digital health with federated learning. NPJ Digital Medicine. **3**(1), 119 (2020) arXiv:2003.08119. https://doi.org/10.1038/s41746-020-00323-1. PMC 7490367. PMID 33015372. S2CID 212747909
31. B. Brik, A. Ksentini, M. Bouaziz, Federated learning for UAVs enabled wireless networks: Use cases, challenges, and open problems. IEEE Access **8**, 53841–53849 (2020)
32. N.H. Tran, W. Bao, A. Zomaya, M.N.H. Nguyen, C.S. Hong, *Federated Learning over Wireless Networks: Optimization Model Design and Analysis* (Proc. IEEE Conf. Comp. Commun, Paris, France, 2019)
33. F. Tacchino, C. Macchiavello, D. Gerace, et al., An artificial neuron implemented on an actual quantum processor. NPJ Quantum Inf. **5**(26) (2019)
34. S.P. Yadav, Vision-based Detection, Tracking and Classification of Vehicles, IEIE Transactions on Smart Processing and Computing, ISSN: 2287–5255. **9**(6), 427–434 (2020). https://doi.org/10.5573/IEIESPC.2020.9.6.427
35. B. Makki, K. Chitti, A. Behravan, M. Alouini, A survey of NOMA: Current status and open research challenges. IEEE Open J. Commun. Soc. **1**, 179–189 (2020)

Federated Machine Learning with Data Mining in Healthcare

Nitesh Singh Bhati, Garvit Chugh, and Bhoopesh Singh Bhati

1 Introduction

With the introduction of machine learning, and artificial intelligence (AI), each and every industry has modernized and is on the verge of booming. One of the by-products of machine learning is data mining, which is helping industries to mass mine the collected data over the past years and make reasonable predictions regarding their own profits. Such methods are providing an unprecedented opportunity for the improvement in big data-driven computational techniques and therefore improving the overall outcome [1].

Healthcare industry has seen a recent boom in the usage of ML-based data-driven techniques in order to find predictions in the patterns of the diseases and how they can be treated or predict how these diseases could mutate. But, this industry faces the most amount of challenges when compared to the other ones, due to the fact that patient history and medical records cannot be shared with anyone without proper permission. For instance, according to acts like Health Insurance Portability and Accountability Act (HIPAA), hospitals can use their own patients' records, but

N. S. Bhati (✉)
Department of Computer Science and Engineering Delhi Technical Campus,
Greater Noida, India

G. Chugh
Department of Computer Science and Engineering, Indian Institute of Technology,
Jodhpur, India
e-mail: chugh.2@iitj.ac.in

B. S. Bhati
Department of Computer Science and Engineering, Chandigarh University, Mohali, India

S. P. Yadav et al. (eds.), *Federated Learning for IoT Applications*, EAI/Springer Innovations in Communication and Computing,
https://doi.org/10.1007/978-3-030-85559-8_15

accessing other hospitals' data is not allowed, and ML-based techniques require a huge amount of raw data for the model to be trained in a robust manner [1, 2].

In such a situation, federated-based machine learning, which provides the ability to work with fragmented sensitive data, is being used. The traditional techniques usually collect data from different parts, join them together, merge them in a single format, and then apply some discovery technique; however, in federated machine learning, it allows a global model based on sharing data with a central server which operates on the data being provided by the different institutes locally [3].

Therefore, federated-based machine learning is the promising future for the healthcare industry in terms of data mining period. The data can be kept private as per the laws of Health Insurance Portability and Accountability Act (HIPAA) and yet be used for training the model, which helps in keeping the patient history private as well. Therefore, the goal of this chapter is to understand how federated machine learning works and how it is being used in the healthcare industry with several applications such as data mining and what kinds of challenges this technique is facing.

1.1 Legal Issues of Patient History

With increase in machine learning usage for prediction of business growth, there has been an increase in the awareness of companies using data for their advantage; therefore, the concerns regarding data privacy and security have been emphasized on, by various non-governmental organizations (NGOs). In response to such requests, there have been formation of commitees by the governments of the nations across the world. One example is the "General Data Protection Regulation" (GDPR) created by the European Union (EU) on May 25, 2018. The main goal of GDPR is to protect private data and the choice to share that private data [4].

After the EU, the United States and China have also come up with various laws to protect private data of the general public. While they allow collection of general public's data, they are under a contract which prohibits them to share this data or exchange it with any third party. Such laws diminish the usage of AI in such scenarios but helps in building a more civil society.

Such statements confirm the fact that these laws do not allow for much AI or ML to be used; however, it must be made clear that using federated machine learning, the intention is not to break any law and implode the right to privacy of the patients, instead understand why the law is so unclear and the implications of them could be very much restrictive, which are blocking the further advancements in the healthcare industry. This fact points out the bigotry of the data science community and how far it can be stretched that distributed or federated learning can contribute a suitable safeguard to the public's privacy [4, 5].

1.2 *Overview of Machine Learning and Data Mining*

Machine learning can be defined as the ability to apply some techniques on raw data to make sense of it and gain some knowledge. Such techniques help in improving the decision-making process with the help of the newly gained information. It is mostly employed in situtations when the data becomes incomprehensible for the human mind, and in order to make machine learning successful, data is mined, in order to convert that raw data into something that can be processed.

Data is collected from various sources, which is an issue in the healthcare industry, as already studied. However, the data in different sources is required to be transformed into one single type of data, only after that, machine learning is possible. This process is known as data mining; it contains various phases, after which the data is transformed. The phases are [6]:

(i) Data cleaning, which adds value in the incomplete data and makes sure there is no redundancy.
(ii) Data integration, which combines different forms of data.
(iii) Data transformation, which works on the outcome of data integration and provides one single format.
(iv) Data reduction, which reduces the attributes of the data (Fig. 1).

1.3 *Different Types of Machine Learning*

Machine learning can be classified in three main categories, supervised, unsupervised, and reinforcement [7].

(i) Supervised learning is a technique which requires a correcting model to identify the correct and incorrect mappings done by the agent between the input examples and the output targets.
(ii) Unsupervised learning is the exact opposite of supervised learning; by eliminating the correcting model, the agent gets to learn on its own if it is mapping the examples correctly or not. It works in a manner of finding patterns which are previously unknown.

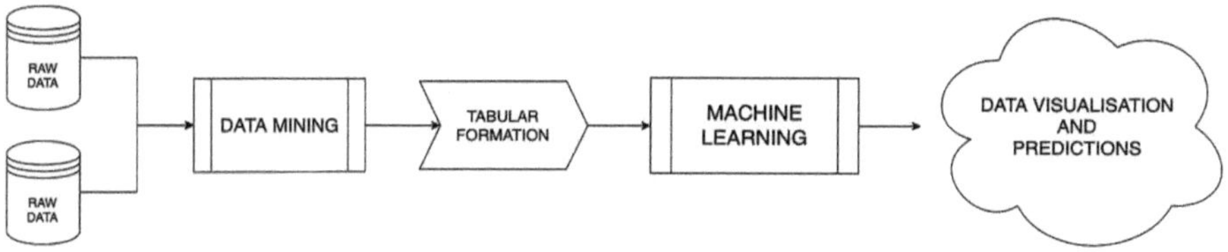

Fig. 1 Machine learning with data mining

Table 1 Most popluar machine learning algorithms

S. no.	ML category	Algorithms
1	Supervised	Support vector machine (SVM), logistic regression, decision tree, random forest, KNN
2	Unsupervised	k-means, Apriori
3	Reinforcement	MDP, Q-learning

(iii) Reinforcement learning is the technique which works on the feedback model of learning; every time an agent makes a correct move, it gets rewarded and therefore tries to maximize the reward by maximizing the correct moves.

Table 1 provides various examples of the most popular algorithms currently being used across the three categories.

2 Federated Machine Learning

Federated machine learning is the technique of using a decentralized manner of working with various sources of data and performing operations on it without using the traditional techniques of data mining. All the operations being performed in federated machine learning are done without the risk of data leakage.

Recent advancements in the field have allowed different types of devices to work in a collaborative manner to provide dataset to this machine learning technique. Hence, the various factors for data optimization are communication cost in massive distribution, unbalanced data distribution, and device reliability. Along with these factors, the fact that the data is being used in an horizontal manner as the data is divided based on device user IDs can also be a factor in data optimization [1, 3, 8] (Fig. 2).

Federated machine learning can be mathematically defined as: let there be N different users $\{U_1, U_2, \ldots, U_N\}$, working on their own sets of data, $\{D_1, D_2, \ldots, D_N\}$. A federated model can be defined as M_F, which does not allow the access of D_1 by any other user (U_i) apart from $i = 1$. The accuract of M_F is A_F, which is very close to the accuracy in terms of the traditional methods of data mining; let it be M_S, which is based on the normal sum of all the datasets, providing no security or privacy whatsoever. In order to define the loss function of federated ML, let Δ be a non-negative number, and if $| A_F - A_S | < \Delta$, then we say that this federated algorithm has Δ – accuracy loss [1, 2, 4].

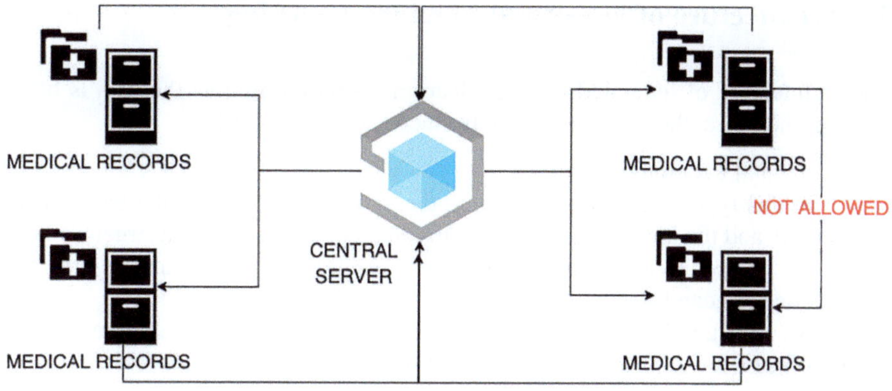

Fig. 2 Federated machine learning process

The federated machine learning problem is based on the following equation (Eq. 1), which helps in solving the empirical risk minimization problem of the algorithm

$$\min_{\Delta \in R} F(\Delta) = \sum_{k=1}^{K} \frac{N_k}{N} F_k(\Delta) \tag{1}$$

where $F_k(\Delta) = \frac{1}{N_k} \sum_{\Delta_i \in D_k} f_i(\Delta)$; w is the model to be learned. The $f_i(\Delta)$ is the loss function. The algorithm of the federated machine learning based on the health industy can be seen in Table 2.

The framework may require extra power of various other machine learning techniques, such as the ensemble technique, which allows to improvise and adapt to the collective power of more than two machine learning techniques for either a faster or a more accurate model of the above representation. There are many wearable devices being used in the real world, which do not require high-powered techniques like SVM; here, lightweight models could be deployed, making it a more realistic model.

Table 2 Algorithm of federated machine learning [9]

Step no.	Actions to be performed
Input:	Different sets of data, $\{D_1, D_2, \ldots, D_N\}$
Output:	A federated model, defined as M_F
1	An initial cloud model is required, which is to be shared with all the users of the network
2	Training of the model M_F using the datasets, not being publically shared, using Eq. 1
3	The different users gets a personalized version of the model to be used locally, which will later be aggregated using $F_k(\Delta) = \frac{1}{N_k} \sum_{\Delta_i \in D_k} f_i(\Delta)$
4	Outcome of step 3 is shared with all the users, and transfer learning is applied, to get the personalized model of each user, which provides the best result of that local machine
5	Repeat the above steps with all the continuously emerging data

3 Architecture of Federated Machine Learning

The architecture of federated machine learning is based on how the data is being divided; therefore, the two types are introduced separately [10].

(i) Horizontal Federated Learning.

In this type of federated learning, the data is separated on the basis of the devices and the identification techniques in that device. The data structure in all of the devices must be the same in order to get this technique to work. The architecture can be seen in Fig. 3.

No leakage of data between databases A, B, and C is allowed. The training process of such a model consists of the following processes:

(a) Local computation of gradient and implementation of an encapsulating algorithm which secures the data and passes it to the server, in a way that only the server is able to decrypt it.
(b) The server makes secure collection of these data packets, and there is complete abstraction in the whole process.

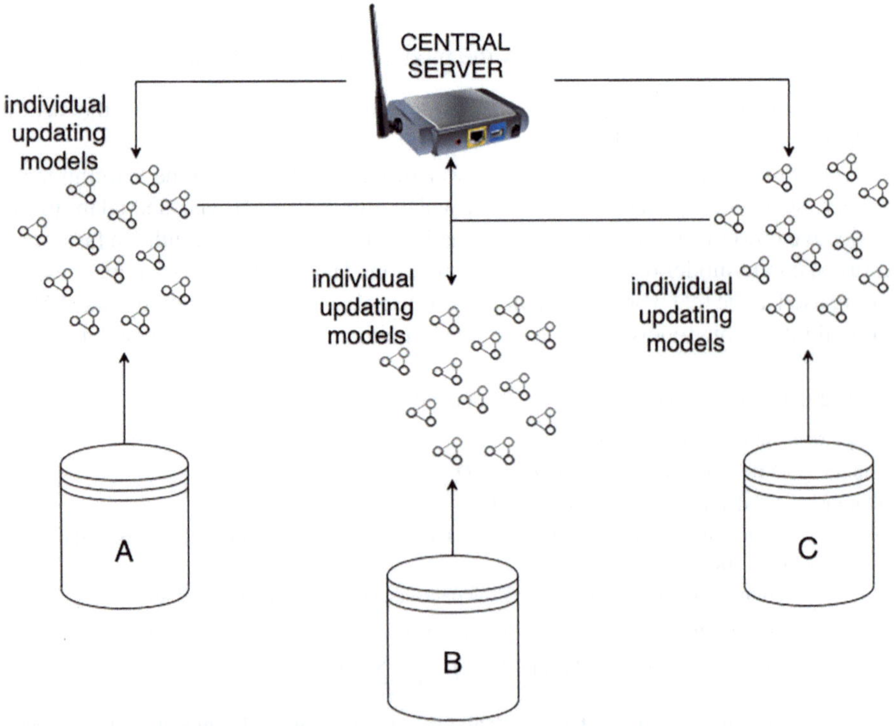

Fig. 3 Horizontal federated architecture

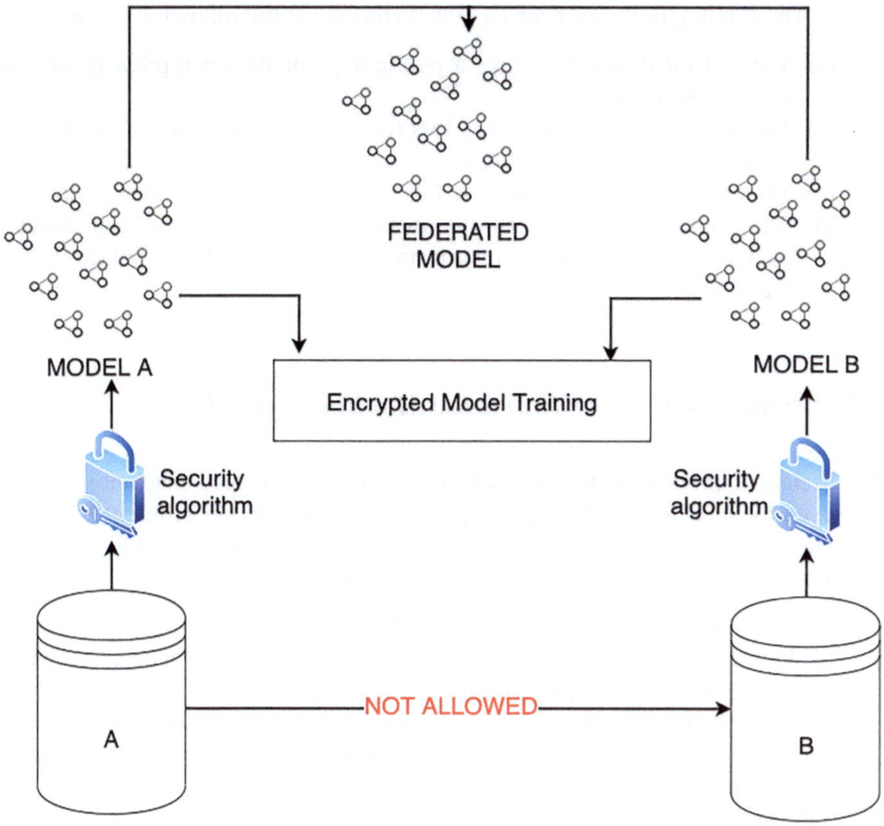

Fig. 4 Vertical federated architecture

(c) The processed results are shared back with the original transmitters in order to update them with the continuously emerging data.
(d) The users update the gradients with the new data.

(ii) Vertical Federated Learning.

In order to maintain privacy in sharing the information between different institutions, let these institutions be A and B as examples, as shown in Fig. 4. The datasets of A and B cannot be directly shared with each other, as mentioned before, there is a security algorithm used which encapsulates the data packets and aggregates the data at a centralized server which performs operations on the data and then shares it back to the individuals which transmitted the data in the first place.

In order to maintain the ethics of security in the centralized server, two properties are kept in check during the process:

(a) Using User-ID as the basis of encryption, this method does not allow the identity of the users to the rest of the cluster.
(b) Encrypted model training is being used in order to make sure the data packets are not revealed to the other users while aggregation of the same.

The training process of such a model consists of the following processes:

(a) The federated model creates a public key for the encryption process to begin at A and B.
(b) The intermediate gradients and loss functions locally generated at A and B are shared with the central server.
(c) The encapsulation process is performed before sending to the server.
(d) The centralized server performs operations on the data and then shares it back to the individuals which transmitted the data in the first place for further updates.

3.1 Privacy Models of Federated Machine Learning

As discussed, privacy is the main concern of using federated machine learning. However, the level of security required by different industries is different according to their needs; therefore, different security models have been created for federated machine learning. In this section, a few of them are discussed.

(i) Secure Multiparty Computation (SMC) Model.
In SMC model, the data providers involved are not aware of any data except for the input and the output. The underlying property of SMC is that users have ZERO knowledge of the intermediate data, but this puts a lot of strain on the algorithms causing heavy computations and high complexity in the algorithm, making it very expensive to use [11].
(ii) Differential Privacy Model.
This model uses noise to block certin attributes of the data in order to prevent the companies to differentiate between the intermediate results, thereby having a secure technique. But this method requires the data to be transferred initially, before any encryption process is applied [12].
(iii) Homomorphic Encryption.
This model uses the technique of parameter exchange during the encryption process. It does not follow the path of SMC or differential privay but somehow has the same disadvantage of having data leaked from the raw collection of data. However, the chances for that are very low, which is why cloud tools use this model for their training [13].

4 Literature Review

In this section, some of the works done in the field of federated machine learning being employed in healthcare industry will be reviewed. Major researchers have given a lot of breakthroughs with the usage of federated machine learning in terms of security and privacy; however, more work is expected in the field, and this review hopes to motivate the readers to pursue in the field.

Chamikara et al. [14] proposed a distributed perturbation algorithm named as DISTPAB, for privacy preservation of horizontally partitioned data. The algorithm removes computational bottlenecks by incorporating techniques of efficient distribution of the task of privacy preservation. Their experiments show that the proposed algorithm, i.e., DISTPAB, is able to yeild high accuracy, efficiency, scalability, and attack resistance.

Huang et al. [15] proposed an algorithm in order to solve the issue of complicated data collection and sharing; they named this algorithm as community-based federated machine learning (CBFL) algorithm. The main aim of this algorithm is to collect data in a clustering fashion, creating some kind of similarity in the non-similar datasets, for instance, the same geographical area is enough to cluster the datasets together. After due experiments, their results show that CBFL is a better version of FL-based learning.

Choudhury et al. [16] proposed a whole new model for privacy in federated learning apart from the traditional methods. Their approach is a syntactic approach while maintaining the standards set by the GDPR and HIPAA. Their experiments provided high throughput with desired level of privacy. However, if this model were to be attacked during the training process, it could become vulnerable.

Chen et al. [17] proposed a solution for wearable technology to provide data for the purpose of federated learning. The framework built by them is known as Fedhealth, which uses transfer learning mode in order to transfer data for sharing in a manner which allows aggregation without compromising security. The experiments performed using Fedhealth have provided enough results to make it usable in a real-world scenario.

Ju et al. [18] proposed a prediction-based algorithm which uses federated learning to make potential stroke-based predictions based on the heath-related data of a patient. Their model adopts in every round and uses the data from each hospital without sharing it with any one them. Therefore its accuracy and privacy, both, are high.

Skripcak et al. [19] proposed a scheme for secure exchange of data within industries without compromising the integrity of the data. However, the framework they proposed was specific to a single use case, which is international research data exchange in the field of radiation therapy and oncology, and their work is dedicated toward development of the said field.

Khan et al. [20] proposed the usage of federated learning in Internet of things (IoT) devices, as these devices are capable of collecting data on their own and transmitting them as well. Their experiment is based on how these IoT devices can use the EDGE network to perform the aforementioned tasks. Their model is based on a reward-incentive system, which pushes the IoT devices to interact more and more (Table 3).

Table 3 Summary of literature review

S. no.	Authors	Domain	Major advantage
1	Chamikara et al. [14]	Efficient algorithm, DISTPAB	High accuracy, efficiency, scalability, and attack resistance
2	Huang et al. [15]	Solution to complicated data sharing	Heavily improved FL learning technique
3	Choudhury et al. [16]	Privacy model	High throughput with desired level of privacy
4	Chen et al. [17]	Fedhealth	Feasibility increased to produce real world scenarios
5	Ju et al. [18]	Prediction-based algorithm	High accuracy and privacy
6	Skripcak et al. [19]	Framework development	Secure exchange of data within industries
7	Khan et al. [20]	IoT-based federated learning	IoT devices can use the EDGE network to perform FL-based tasks

5 Limitations and Solutions

Traditionally, federated average is used to train the federated model, which is the ability to work with a non-similar dataset by forcing the clients to work with one model, same for everyone. However, there are certain limitations with this naïve way, as it does not work with a set of strongly skewed database [21].

For such a problem, there are two solutions that have been proposed:

(i) Consensus Solution.

The issue in federated model is that it uses a training technique based on the aggregated datasets, with the goal of minimized loss w.r.t. uniform distribution, but for most cases, the distribution is not uniform enough. In order to solve this problem, the data is being forced to change its attributes in order to get a complete uniform dataset.

During the testing phase of this approach, it has been found that it works impeccably in a small collection of data. Therefore, other approaches have also been uncovered to perform the operation at a large scale. For instance, a q-Fair learning technique, when applied with federated learning, reduces variance in datasets, which helps improve the flexibility and scalability power of federated machine learning.

(ii) Pluralistic Solution.

A consensus solution is not possible for every component of a dataset, as it is forcing the dataset to change into something it is not. Therefore, there's a need for evolution in the process. Upon further research, it was found that the hetergenity property of the consensus solution can be used to deal with different datasets and even draw data from these sources using a single method.

Such a solution is multitask learning; it has the ability to directly find relationships between non-similar data points and create a single global model. However, this can only be achieved when the data shares some kind of similarity in the data structure, i.e., sharing sparsity, sharing low-rank structure, and graph-based relatedness.

6 Conclusion

As of now, a mix of guidelines and morals makes it hard to share medical or health-related information in any manner, for any kind of purposes. The legal issues are identified with the legitimate reason for privacy and anonymization. In particular, there has been a hesitance toward the sharing of data because in the past, there have been public leaks of personal data leading to various kinds of harassments taking place with the patients. An answer to this problem could be to permit scientists to train their AI programs without the information ever leaving the centers; this phenomena is termed as federated learning. This protected practice makes it conceivable to gain infromation from clinics and can be applied across different machine learning-based models. However, the current technology is still not enough to make sufficient impact for accurate and robust predictions as the data pool is still limited due to limitations of federated learning. A solution for this situation might be the usage of blockchain; in the future, an integrated design of blockchain with federated learning might facilitate the distributed learning aspect of federated learning.

References

1. F. Zerka, S. Barakat, S. Walsh, M. Bogowicz, R.T. Leijenaar, A. Jochems, et al., Systematic review of privacy-preserving distributed machine learning from federated databases in health care. JCO Clin. Cancer Informatics **4**, 184–200 (2020)
2. Q. Yang, Y. Liu, T. Chen, Y. Tong, Federated machine learning: Concept and applications. ACM Transactions Intelligent Systems and Technology (TIST) **10**(2), 1–19 (2019)
3. J. Xu, B.S. Glicksberg, C. Su, P. Walker, J. Bian, F. Wang, Federated learning for healthcare informatics. J. Healthcare Informatics Res. **5**(1), 1–19 (2021)
4. F. Zerka, S. Barakat, S. Walsh, M. Bogowicz, R.T. Leijenaar, A. Jochems, et al., Systematic review of privacy-preserving distributed machine learning from federated databases in health care. JCO Clin. Cancer Informatics **4**, 184–200 (2020)
5. Q. Li, Z. Wen, Z. Wu, S. Hu, N. Wang, B. He, A survey on federated learning systems: Vision, hype and reality for data privacy and protection. arXiv preprint arXiv:190709693 (2019)
6. B.S. Bhati, G. Chugh, F. Al-Turjman, N.S. Bhati, An improved ensemble based intrusion detection technique using XGBoost. Trans. Emerg. Telecommun. Technol., e4076 (2020)
7. N.S. Bhati, M. Khari, V. Garcia-Diaz, E. Verdu, A review on intrusion detection systems and techniques. Inter. J. Uncertainty Fuzziness Knowledge-Based Syst. **28**(Supp02), 65–91 (2020)
8. N. Rieke, J. Hancox, W. Li, F. Milletari, H.R. Roth, S. Albarqouni, et al., The future of digital health with federated learning. NPJ Digital Med. **3**(1), 1–7 (2020)

9. G.A. Kaissis, M.R. Makowski, D. Rückert, R.F. Braren, Secure, privacy-preserving and federated machine learning in medical imaging. Nat. Machine Intelligence **2**(6), 305–311 (2020)
10. K. Bonawitz, H. Eichner, W. Grieskamp, D. Huba, A. Ingerman, V. Ivanov, et al., Towards federated learning at scale: System design. arXiv preprint arXiv:190201046 (2019)
11. T. Li, A.K. Sahu, A. Talwalkar, V. Smith, Federated learning: Challenges, methods, and future directions. IEEE Signal Process. Mag. **37**(3), 50–60 (2020)
12. R.C. Geyer, T. Klein, M. Nabi, Differentially private federated learning: A client level perspective. arXiv preprint arXiv:171207557 (2017)
13. B.S. Bhati, C.S. Rai, Analysis of support vector machine-based intrusion detection techniques. Arab. J. Sci. Eng. **45**(4), 2371–2383 (2020). https://doi.org/10.1007/s13369-019-03970-z
14. S.P. Yadav, Emotion recognition model based on facial expressions. Multimed. Tools Appl. (2021). https://doi.org/10.1007/s11042-021-10962-5
15. L. Huang, A.L. Shea, H. Qian, A. Masurkar, H. Deng, D. Liu, Patient clustering improves efficiency of federated machine learning to predict mortality and hospital stay time using distributed electronic medical records. J. Biomed. Inform. **99**, 103291 (2019)
16. S.P. Yadav, K.K. Agrawal, B.S. Bhati, et al., Blockchain-Based Cryptocurrency Regulation: An Overview. Comput. Econ. (2020). https://doi.org/10.1007/s10614-020-10050-0
17. Y. Chen, X. Qin, J. Wang, C. Yu, W. Gao, Fedhealth: A federated transfer learning framework for wearable healthcare. IEEE Intell. Syst. **35**(4), 83–93 (2020)
18. C. Ju, R. Zhao, J. Sun, X. Wei, B. Zhao, Y. Liu, Y. Jin, Privacy-preserving technology to help millions of people: Federated prediction model for stroke prevention. arXiv preprint arXiv:200610517 (2020)
19. T. Skripcak, C. Belka, W. Bosch, C. Brink, T. Brunner, V. Budach, et al., Creating a data exchange strategy for radiotherapy research: Towards federated databases and anonymised public datasets. Radiother. Oncol. **113**(3), 303–309 (2014)
20. L.U. Khan, S.R. Pandey, N.H. Tran, W. Saad, Z. Han, M.N. Nguyen, C.S. Hong, Federated learning for edge networks: Resource optimization and incentive mechanism. IEEE Commun. Mag. **58**(10), 88–93 (2020)
21. C. Fung, C.J. Yoon, I. Beschastnikh, The limitations of federated learning in Sybil settings, in *23rd International Symposium on Research in Attacks, Intrusions and Defenses*, (RAID, 2020), pp. 301–316

Federated Learning for Data Mining in Healthcare

Shivani Sharma, Akash Kesarwani, Shreyshi Maheshwari, and Bipin Kumar Rai

1 Introduction

Because increasingly more information is becoming promptly available from numerous sources, such as clinical offices, patients, protection suppliers, and drug enterprises, there has been a flood of interest in medical services information examination as of late. This presents a unique potential for the headway of scientific procedures to uncover information driven bits of knowledge for improving consideration conveyance quality [11]. Because of the staggering nature of the structure and techniques of medical services, information about medical services is regularly dispersed. Various medical clinics, for instance, may approach similar information, but only their own patient populaces' health records are kept. People's protected health information (PHI) is remembered for these reports, which are very delicate. To control the way of obtaining and investigating such information, exacting guidelines have been developed, for example, the Health Insurance Portability and Accountability Act (HIPAA) [1]. Current data mining and AI (machine learning) innovations, for example, deep learning [6], which generally include a lot of preparing information, face a huge test. Combined learning is a worldview that has as of late acquired a large foothold because it has extensive potential for learning with divided sensitive information. Instead of consolidating information from different sources or depending on the ordinary revelation and replication approach, it permits clients to prepare a common worldwide model with a focal worker while holding the information in the organizations where they were gathered. "United learning," "unified learning," and "federated learning" are other used phrases. Patrick Hill, a logic educator, made the Federated Learning Community (FLC) in 1976 to unite individuals to concentrate together, which assisted understudies with mitigating the

S. Sharma · A. Kesarwani · S. Maheshwari · B. K. Rai (✉)
Department of IT, ABES Institute of Technology, Ghaziabad, Uttar Pradesh, India

S. P. Yadav et al. (eds.), *Federated Learning for IoT Applications*, EAI/Springer Innovations in Communication and Computing,
https://doi.org/10.1007/978-3-030-85559-8_16

secrecy and isolation that they encountered everywhere in research colleges [2]. Following that, numerous endeavors were undertaken to set up organizations of information and information databases. In 2005, Rehak et al. [11] set up a reference model for building an interoperable vault engineering through combining stores, in which metadata is gathered from participating storehouses and maintained in a focal library with a specific access requirements. This current model's definitive objective is to stimulate gaining from an assortment of data archives. These combined learning network or united inquiry administration exercises have been helpful when planning unified learning calculations. Regarding medical care information examination, unified learning holds a great deal of guarantee. The sensitive patient information should remain in neighborhood organizations or with singular purchasers without going out during the alliance for both providers (e.g., establish a model to predict readmission risks at medical clinics with electronic health records (EHRs) of patients [10]) and shopper (patient) based applications (e.g., screening(atrial fibrillation with electrocardiograms captured by smart devices). The purpose of this chapter is to review the essentials of combined learning, discuss broad arrangements and issues, and envision how it could be utilized in healthcare. Following an efficient examination of unified learning, the key issues and recent advancements are summarized in this investigation. By discussing a productive recent trial, we show the capability of combined learning strategies in healthcare. Finally, we go over the key prospects and unanswered inquiries for future medical service applications.

Differences from Previous Reviews Recently, there have been some review papers on federated learning, Yang et al. [15], for example, compiled an early federated learning survey that summarizes the various privacy preservation approaches that)can be used in federated learning. Personalization-strategies [5], semi-supervised learning algorithms [3], hazard-models [9], and mobile-edge-networks [8] are some-of the sub-problems of federated learning that have been studied by some researchers. Kairouz et al. [4] reviewed recent progress and addressed a long list of unsolved issues and challenges. Li et al. [7] published a study on federated learning from a machine perspective.

2 Federated Learning

Federated learning involves utilizing a focal worker to prepare a first-rate shared worldwide model from decentralized data dispersed through countless various customers (Fig. 1). Expect there are K enacted customers where the information is stored numerically (a customer could-be a mobile phone, a portable gadget, or a clinical facility information database, and so on); allow D_k to represent the information conveyance for customer k and

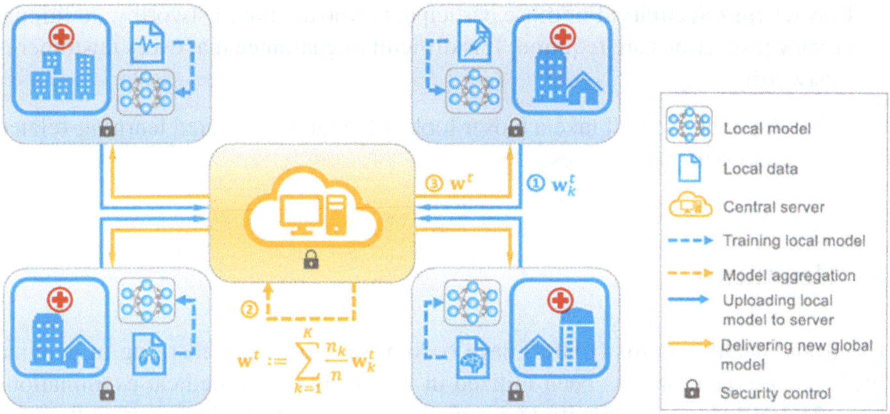

Fig. 1 The federated learning system is shown in this diagram.The model is trained in a distributed way:-institutions regularly report local updates on central server in order to learn a global model; the central server aggregates the updates and sends back modified global model's parameters [42]

n_k the quantity of tests the customer has open. The complete example size is indicated by the letter n. The federated learning issue is fundamentally an issue of experimental danger minimization of the structure [30, 31]:

$$\min F(w) := \sum_{k}^{k=0} \frac{n_k}{n} F_k(w) \, where \, F_k(w) := \frac{1}{n_k} \sum_{x_i \in D_k} f_i(w)$$

where w is the to-be-learned model parameter. A loss function is used to define the function f_i, which is based on a pair of input-output data pairs-x_i,-y_i.-Usually x_i Rd and y_i R or y_i 1, 1 are used. Here are some simple examples:

- Linear regression $f_i(w) = \frac{1}{2}(x_i w - y_i)^2, y_i \in R$;
- Logistic regression $f_i(w) = - log(1 + exp(-y_i x_i w)), y_i \in \{-1, 1\}$;
- Support vector machine $f_i(w) = max\{0, 1 - y_i x_i w\}, y_i \in \{-1, 1\}$;

In particular, there are many challenges to federated learning algorithms [], specifically:

- **Statistical Challenge:** *ExiDk [fi(w; xi)] = ExiDk [fi(w; xi)] = ExiDk [fi(w; xi)] = ExiDk [fi(w; xi)] = ExiDk [fi(w; xi)] = ExiDk [fi(w; xi)] = ExiDk [fi(−w; xi)] = Ex* Therefore, it is unlikely that local data points are indicative of overall distribution.,i.e., *ExiDk [fi(w; xi)] = F(w)*.
- **Communication Efficiency:** The number of K clients is high, and it may be much higher than the average number of training-samples-stored in enabled clients, i.e., K>> (n/K).

- **Privacy and Security:** For those participants who are not trustworthy, additional privacy protections are required. It is difficult to guarantee that every customer is trustworthy.

Following that, we will take a closer look at the latest federated learning-related work on addressing such issues.

3 Healthcare

Electronic health records (EHRs) have become an amazing wellspring of genuine medical care data that has been utilized in various basic biomedical examinations [12, 13]. EHRs include primary and irregular predispositions that obstruct the generalizability of results, regardless of the way that they present a huge measure of patient data for audit. Obermeyer et al. [14] found that a broadly utilized calculation for deciding enlistment, specifically health administrations, was slanted toward African Americans, putting heathier Caucasian patients at a similar danger level. In view of errors in hidden admittance to mid or low portrayal in preparing information, these improperly tuned calculations can occur for an assortment of reasons. Unquestionably, gains from EHR data that is more knowledgeable about the world's population and stretches beyond a clinic or isolated area is an approach to moderate the danger of such one-sided calculations. Surprisingly, because of different variables such as divergent information frameworks and security issues, it is dubious that information could at any point be connected together in a typical data set from which to learn. The turn of events and utilization of organized basic data models, for example, OMOP [15], take into account more diverse replication questions; however, it leaves the problem of shared knowledge access unaddressed. Unified learning can be the instrument to enable ML of EHR data. Combined learning is a practical technique for associating EHR data from clinical facilities, permitting them to share their encounters as opposed to their data while maintaining confidentiality [16–21]. In these cases, iterative changes in gaining from wide and complex clinical informational indexes would significantly expand the effectiveness of the ML model. Patient similarity learning [22], patient representation learning, phenotyping [23, 24] and prescient demonstrating [19, 25, 26] are some of the areas that have been concentrated on in a combined learning climate in medical care. Lee et al. [22] proposed a protection safeguarding stage for patient likeness learning through organizations in a unified climate. Without trading patient-level details, their model will discover comparable patients starting with one emergency clinic then onto the next. In a unified learning climate, Kim et al. [23] utilized tensor factorization models to change huge electronic health records into solid aggregates for information examination, used a united way to perform both patient representation learning and determined comorbidity phenotyping [24], with positive outcomes. Vepakomma et al. [28] created a scope of arrangements dependent on an appropriated deep learning approach called Split NN [27] to empower health associations to work together on deep learning model preparation without sharing delicate raw data or model data.

4 Data-Mining-Approach

In prosperity checking-systems,-data mining is used to extract the information from low-level sensor data and associate it with a higher-level data representation. Thus, ongoing health monitoring frameworks have prioritized the information assortment step to retrieve more helpful information dependent on master buyer prerequisites. Data mining methods utilized in health global positioning frameworks with wearable sensor information have shifted, and it is not unusual for all procedures to be utilized in a similar design. The most well-known methodologies utilized for wearable sensor information to give helpful data, for example, the assignments portrayed in sect. "Federated Learning", are laid out in this segment. The most well-known and commonly utilized way to deal with mining data from wearable sensors, paying little mind to the data mining method utilized are shown in Fig. 2.

When it comes to data mining, raw sensor data is often used as a starting point, as demonstrated in the diagram. Both training and testing purposes used sensor data. Training data is utilized to become familiar with the system and create a working model, while testing data is utilized to make a model of real-world use. This data mining method is proposed as an overall stream data mining solution, both-supervised-and-unsupervised-with the objective of finishing any data mining mission. The data mining technique's key advances are as follows:

1. Information pre-processing.
2. Extraction and selection of features.
3. To perform tasks such as recognition, prediction, and decision-making, data modeling is used to learn the input features (while taking into account expert knowledge and metadata).

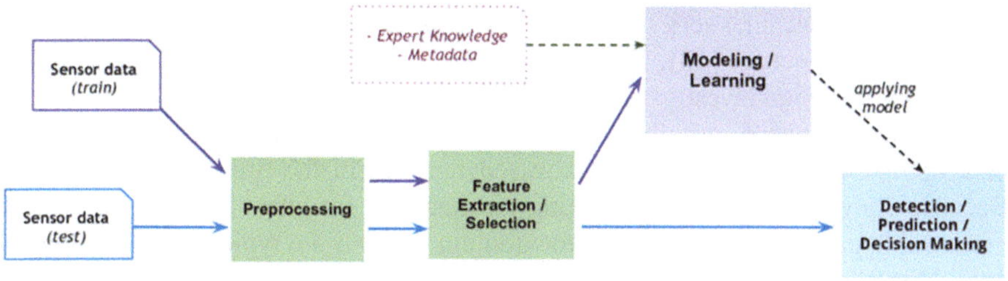

Fig. 2 The main data mining method for wearable sensor data is represented by a generic architecture [20]

Expert ability, historical information measures, electronic health records, and stable boundaries are largely fundamental boundaries to consider when utilizing data mining and AI methods. This metadata takes into account more relevant examination and fortifies the data extraction measure [20]. For instance, any medical services framework that utilizes HR sensor information should examine the impact of metadata, including age,-sex,-weight, and prescription on significant reasoning to find, for example, simple erratic pulses or to tailor vital heartbeats based on metadata [43]. Following that, we will go over the main advancements in data mining technique in detail:

4.1 Preprocessing

To remove frequency noise from sensor signals, other frequency-domain methods such as power spectral density (PSD), fast Fourier transforms (FFT), and low pass/high pass filtering tools are commonly used [42, 50]. When data is collected from several wearable sensors [39], it must be formatted, normalized, and synchronized. Preprocessing of the raw data is required in any wearable sensor network because of the presence of noise, unusual movement, and sensor oversights. In the medical care space, preprocessing involves (1) eliminating objects from odd data and (2) eliminating frequency noise [38, 39]. As per the writing, fabricated modules have normally utilized edge-based strategies to channel sensor data [36, 37] or measurable apparatuses to introduce missing information focuses for sifting objects [40]. Extraction/selection of features is a method of planning and developing a model to extract useful data from large and real-world datasets, which is usually reflected in the raw holding data mining approach. The goal of feature extraction is to discover them in data set characteristics that are exact replicas of the original data [66]. Feature extraction, particularly in wearable sensor data, gives a significant portrayal of the sensor information that can form the connection of raw data with expected data [53] dependent on the size and intricacy of the raw data. Furthermore, in the element extraction and highlight decision cycles, decreasing the measure of sensor network information is another way, prompting a coordinated vector of highlights as a contribution to data mining strategies such as classifier techniques [37]. Because most wearable sensor information used to follow essential sign boundaries is captured as persistent time series readings, most of the highlights considered are time series signal properties [63]. The time domain and phantom space are the two significant parts of sign investigation. Basic waveform characters and statistical parameters identified with discernible attributes in data streams, such as mean, variance, pick counts, etc., are commonly extracted features in the time domain [36, 40]. Because conventional decision-making mechanisms for vital signs are focused on measurable patterns in the signal, time domain features are typical in physiological data. Nonetheless, to acquire a more profound comprehension of the intermittent conduct of time series, clinical examination has zeroed in on highlight obtained from recurrence spaces. Table 1 summarizes the most regularly referred to

Table 1 A summary of recent work in healthcare federated learning

Problem	ML Method	No. Of clients	Data
Patient similarity learning [22]	Hashing	3	MIMIC-III [3]
Patient similarity learning	Hashing	20	MIMIC-III
Phenotyping [23]	TF	1-5	MIMIC-III, UCSD
Phenotyping [24]	NLP	10	MIMIC-III
Representation learning [28]	PCA	10-100	ADNI, UK Biobank, PPMI, MIRIAD
Mortality prediction [19]	Autoencoder	5-50	eICU Collaborative Research Database
Hospitalization prediction [25]	SVM	5-10	Boston Medical Center
Preterm birth prediction [16]	RNN	(50	Cerner Health Facts
Mortality prediction [29]	LR, NN	31	eICU Collaborative Research Database
Mortality prediction [35]	LR, MLP	2	MIMIC-III
Activity recognition	CNN	5	UCI Smartphone
Adverse drug reactions Prediction	SVM, MLP, LR	10	LCED, MIMIC
Arrhythmia detection	NN	16,32,64	PhysioNet Dataset
Disease prediction	NN	5,10	Pima Indians Diabetes Dataset [95], Cleveland Heart Disease Database
Imaging data analysis	VAE	4	MNIST, Brain Imaging Data
Mortality prediction	LRR, MLP, LASSO	5	Mount Sinai COVID-19 Dataset

highlights obtained from wearable sensor information in the writing. While Bsoul et al. proposed many highlights for examining ECG flags; the fundamental focal point of examination is on R focuses and their properties in-ECG-heartbeats such as R-R stretches [53]. There are also heuristic and specific field highlights to remember in addition to the highlights contained in these areas [60]. Feature selection is an efficient solution for selecting more discriminative features. Feature selection determination techniques normally recognize a subset of high-dimensional information that is less relevant and adds to the student's yield [37]. Utilizing feature selection strategies on physiological data decreases the dimension of the data [47], which measurably selects the subset of the main feature selection [34, 47]. In the medical field, there are three popular approaches to dimension reduction. Other feature selection methods mentioned in the literature include-threshold-based rules [45], analysis of variance (ANOVA) [39], and Fourier transforms [40]. Despite the fact that most proposed systems in the medical services space have feature extraction/

selection stages, the key challenge remains balancing the best feature extraction/selection process with the system and costs. Because modeling techniques should deal with the raw extracted features, using feature selection in real-time systems would be expensive. The method must deal with redundant data in these situations, lowering the-accuracy-of the results. The above-mentioned problem is inextricably linked to contemplation.

1. In the framework, the health parameters are selected from health monitoring system.
2. The aim of the health monitoring system is to perform data mining tasks. Recent science, on the other hand, has yet to discuss a solution.

4.2 Support Vector Machine

When modeling-and learning techniques are used, they can collect a lot of data. Furthermore, as multiple sensors are utilized, the information becomes multivariate and potentially dependent. This means that sufficient data processing techniques are needed to make sense of the data. The most popular algorithms used for wearable sensor data are described in this section. Each algorithm is depicted in specialized detail, alongside the most significant instances of how it has been utilized in medical care. Likewise, each technique convenience, execution, and difficulties in the clinical area are referenced. Because the SVM can handle high-dimensional data with a restricted arrangement of preparing highlights, it has become common for mining physiological data in clinical applications. SVM techniques take into account common health parameters such as ECG, HR, and SpO2, which are typically used in a short-term and annotated format. Hu et al. [44] applied SVM to ECG signals in order to detect arrhythmia. The method and input are the pulse's reduced power spectral density properties. The sensitivity, precision, and accuracy of the results are used to determine the method's efficiency. In healthcare settings, anomaly identification and decision-making tasks are often proposed using SVM techniques. SVM, on the other hand, is not an effective method for incorporating main information in order to use metadata or symbolic awareness in combination with sensor measurements. SVM was also unable to detect the unexpected in unlabeled results, unlike other classifiers.

4.3 Neural Networks

A neural network (NN) is a form of artificial intelligence that is used to classify and predict data [46]. The NN approach changes the network weights for subsequent learning iterations by analyzing the records' known classifications and comparing them to expected classes. NN has also been extended to multi-sensor networks in order to manage the sophisticated analysis of multi-variety data [47]. The efficiency of PPG pulses was estimated using a multi-layer perceptron (MLP) neural network. As a result, NN is not thought of as a strategy that can handle a large variety of data sets. This network optimizes the number of hidden layer nodes (2–20) in validation iterations by taking many individual signal quality metrics as input. Consequently, the result falls into two signal quality categories. It used a variety of indices to assess the work, including sensitivity, specificity, and accuracy. Using ECG and accelerometer sensors, Vu et al. [37] introduced a method for understanding heart rate variability (HRV) patterns. A three-layer neural network was used to gradually learn and classify the extracted patterns in this process. Three data classifications for location, process, and heart status were defined by three output layer nodes. The replicator neural network (RNN) is another kind of NN [49]. To summarize, because learning improvement in NN is complex, the approach is widely used for clinical decision-making management and a complicated data set modeling process is a black box process. However, similar to SVM, it was unable to use domain information to improve the outcomes. Furthermore, because the NN approach must explain each input data, the decision tree approach is one of the most important learning strategies for representing rule classification efficiently. Using this method, which involves constructing a tree-like mode, the most robust features have been identified for initial splitting of the input data. Decision trees are a valid method to use to make the best decision in various areas of the medical domain. When managing perplexing and noisy information, the C4.5 calculation [50] is utilized, which ascertains the underlying hub mistake rate and prunes the tree to make a more effective sub-tree when utilizing transient information with few subjects, and decision trees have received more consideration in the clinical area. Because of the making of free levels in the decision tree, this methodology is additionally ideal for overseeing multivariate sensors. Frantzidis et al. [50, 53] proposed another movement acknowledgment framework. It separated positive and negative feelings of subjects when seeing emotional pictures utilizing the C4.5 decision tree calculation dependent on the Mahalanobis gap. The tree is constructed utilizing biosensor information highlighted in two layers of feeling segregation. To settle on a legitimate choice, the applied classifier utilized a standard based strategy. The assessment of warmth stress hazard is another order movement [51] that decision trees are utilized for. To create a C4.5 tree, this technique used information boundaries such as mean skin temperature, cooling incitation, and encompassing temperature, as well as certain principles. The tree's effectiveness is classified as a solid threat or protection label in 95% of cases. Bellos et al. [35, 46] discuss another possible use of decision trees in this context. A decision emotionally supportive network is used in [52] to arrange the seriousness of health level utilizing random-forest (RF) characterization as a variation of the decision tree. This methodology joined the greatness of persistent illness for ECG and HR information utilizing a combination of designs. Another subset of highlights was picked for the development of each decision tree. The framework utilized limit-based principles to pick the best tree. The framework's exactness was estimated against some foreordained objectives. In [50], other decision tree executions in medical services frameworks are specified. All in all, decision tree strategies are confined to the space of the model's built highlights as inputs.

As an outcome, extricating secret information from choked highlights will be unrecognizable. In addition, because the quantity of highlights can influence the strategy's presentation, decision tree models are only occasionally utilized for broad and complex physiological data. Notwithstanding, a decision tree arrangement has been utilized in the writing to tackle the entirety of the errands recorded in section "Healthcare" because it is quick and simple to carry out.

4.4 Gaussian Mixture Models

Another numerical model for arrangement and example acknowledgment is the Gaussian mixture model (GMM) [54]. The information is attempted to be a straight blend of Gaussian dispersions in this interaction. Studies that use GMM usually use explained clinical data to assess the model's efficiency. To have safe body sensor correspondences, Wang et al. [56] suggested a GMM strategy based on inter pulse intervals (IPI) of ECG. This instrument is centered on the way that every individual's ECG signal conduct is unique, and it tends to be utilized as a mark to check different subtleties (e.g., prescription conveyance content data). Clifton et al. [56] proposed a methodical approach to dealing with the Gaussian technique in order to increase the ability to tailor patient monitoring. The Gaussian strategy was utilized in this structure to effectively gauge the circulation of physiological data values. The Gaussian interaction's mean capacity is then used to distinguish unusual SpO2. The strategy was made to assist with the evacuation of items and missing information from singular subjects. Another GMM-based examination is presented in [55] and recommends utilizing ECG signs to analyze coronary vein disease conditions. This framework grouped sign scenes into ordinary and constant states utilizing numerous arrangement techniques (e.g., SVM, NN, and GMM). The precision of the GMM classifier was discovered to be exceptional in this analysis. Given the way that GMM can recognize obscure data in physiological information, it has only been utilized in a couple of forecast assignments. Applying GMM progressively in systems is not doable because of the long calculation time expected to assemble the models. Moreover, because of the idea of health boundaries in genuine conditions, the introduction stage in GMM does not generally advance demonstrating (as GMM is drawn nearer as a blend of Gaussian differentiation).

4.5 Hidden Markov Models

The hidden-Markov-model (HMM) is a probabilistic model for displaying consecutive intelligence [] centered on a histogram of subsequent state probabilities, seen as a Markov chain, to decide the probability of each state occurring. Signs are usually displayed with the help of Markov chain using the probability of registered proportion [64]. Another HMM study dependent on distinguishing outsider portions in wearable sensor time series themes was presented in [64]. ECG and accelerometer sensor information were utilized to evaluate the model's exactness. The signs were displayed utilizing Markov chains, with the probability of each portion registered. The striking sections are those with the least probabilities (only fascinating segments). HMM is predominantly utilized for peculiarity location and not for some other errands. Moreover, this methodology apparently has not been extended to the investigation of different health boundaries or big data collections. Notwithstanding, in view of HMM's capacity to display information's unusual activities, altered renditions of HMM can be applied to a more extensive scope of issues in the medical services field.

4.6 *Rule-Based Methods*

RBR is a vital technique for detecting patterns, anomalies, and one-of-a-kind events that are based on rules and conditions that have been predefined and saved. The RBR process is required to be applied on any wearable sensor device [57] because the fundamental task of health monitoring is to track down the obvious issue in physiological data [57]. Using ECG sensor data, a standard-based technique was developed for identifying various forms of arrhythmias. This chapter presents a collection of threshold-based rules that can be used to differentiate between different arrhythmias using ECG features. Obviously, the expert scheme provides the definitions of the conditions (range of values) for different features in ECG signals. Another study [58] dealt with the general anesthetic issue. The aim of this project is to keep track of different trends requested by clinicians. Rule-based approaches can detect anomalies and find unusual trends in unlabeled data sets because they rely on a set of rules provided by experts or domain experts. In order to learn the features and actions of the data, these methods, on the other hand, are biased against predefined rules. For decision-making and prediction, rule-based methods, as well as a customized framework with context-aware objectives, are inadequate. Furthermore, they were unable to manage complex physiological data in real-world experiments with unpredictably patterned data.

4.7 *Statistical Tools*

In addition to the procedures depicted above, for data mining techniques,-other general statistical parameters and probability-functions have been used in the literature [59]. In the majority of health monitoring systems research [61], simple statistical parameters (e.g., mean, variance, etc.) or statistical functions (e.g., risk function [96], factor analysis [60]) will suffice to formulate the data and obtain the expected information. This type of information investigation is broadly used to improve on information highlights for model development in multi-sensor networks [62]. Statistical tools are often used to model input physiological data using statistical features and behaviors of health parameters using statistical tools. As a result, these models are capable of detecting anomalies that are difficult to detect even by experts. Domain information and background metadata, however, cannot be incorporated into statistical models. Given the fact that these methods are unsuitable for making decisions, they have been used in clinical settings with transient data to monitor the most important changes in the data stream.

4.8 Frequency Domain/Wavelet Analysis

The frequency domain investigation of time series information allows for the extraction of previously unseen patterns and trends through data [62]. The frequency domain zeroing raises significant data features, particularly for wearable sensor time series data, similar to statistical tools. In this study, Fourier and wavelet transform analyses have primarily been used for-feature selection and other tasks in physiological data, such as vital sign estimation [57] and time series filtering and physiological data noise reduction. Frequency domain analysis can detect non-obvious anomalies as well as model and discover complex patterns in data. However, since these methods have long calculation times, they are not suitable for use in real-time decision-making systems. Moreover, because of an absence of models, these models have not been widely used for prediction tasks.

5 Applications

This selection demonstrates measures that take into consideration effective AI while safeguarding data and lawful authorization through various gatherings or registering hubs. Virtual console prediction [32], smart retail, account [10], and vehicle-to-vehicle contact are some encouraging models that represent these capacities. We will focus on applications in the medical services space in this segment, and we will likewise address promising applications in different areas in light of the fact that specific ideas can be applied to medical care.

6 Conclusion

Other data mining procedures, which are generally utilized in physiological data examination, are among the techniques thought of. Other examples include Bayesian networks for classifying stress levels [51, 63] or predicting patient states [47], Fuzzy state machines for assessing health status and providing warnings [48], logistic regression models [63], and interaction rules [50] for anticipating feelings of anxiety and hospitalization in hemodialysis patients, separately. The utilization of data mining methods is comparable to health boundaries assembled with wearable sensors, as indicated by the overview. As can be seen, the most unpredictable health boundaries are the ECG, HR, SpO2, and RR, for which scientists have utilized an assortment of data mining strategies to look at these fundamental signs. SVM, NN, decision tree, and recurrence/wavelet are broadly utilized data mining strategies for wearable sensor data handling, but they have only been seen on the ECG health boundary. It has likewise shown designs in the data investigation techniques used to follow health boundaries such as ECG, RR, HR, BP, and BG. Finally, the arising challenges of utilizing data examination procedures in health monitoring frameworks were met with exceptional attention. As an outcome, this chapter discusses various medical care administration alternatives by considering various components.

References

1. L.O. Gostin, National health information privacy: regulations under the health insurance portability and accountability act. JAMA **285**(23), 3015–3021 (2001)
2. P. Hill, *The Rationale for Learning Communities and Learning Community Models* (ERIC, 1985)
3. Y. Jin, X. Wei, Y. Liu, Q. Yang, A survey towards federated semi-supervised learning. arXiv:2002.11545 50 (2020)
4. A.E. Johnson, T.J. Pollard, L. Shen, H.L. Li-wei, M. Feng, M. Ghassemi, B. Moody, P. Szolovits, L.A. Celi, R.G. Mark, Mimic-iii, a freely accessible critical care database. Sci Data **3**, 160035 (2016)
5. P. Kairouz, H.B. McMahan, B. Avent, A. Bellet, M. Bennis, A.N. Bhagoji, K. Bonawitz, Z. Charles, G. Cormode, R. Cummings, et al., Advances and open problems in federated learning. arXiv:1912.04977 (2019)
6. V. Kulkarni, M. Kulkarni, A. Pant, Survey of personalization techniques for federated learning. arXiv:2003.08673 (2020)
7. Y. LeCun, Y. Bengio, G. Hinton, Deep learning. Nature **521**(7553), 436–444 (2015)
8. T. Li, A.K. Sahu, M. Zaheer, M. Sanjabi, A. Talwalkar, V. Smith, Federated optimization for heterogeneous networks. arXiv:1812.06127 (2019)
9. W.Y.B. Lim, N.C. Luong, D.T. Hoang, Y. Jiao, Y.C. Liang, Q. Yang, D. Niyato, C. Miao, Federated learning in mobile edge networks; a comprehensive survey. arXiv.1909.11875 (2019)
10. J. Xu, F. Wang, Federated learning for healthcare informatics. arXiv:1911.06270[] (2019)
11. D. Rehak, P. Dodds, L. Lannom, A model and infrastructure for federated learning content repositories, in *Interoperability of web-based educational systems workshop*, vol. 143, (Citeseer, 2005)
12. B.S. Glicksberg, K.W. Johnson, J.T. Dudley, The next generation of precision medicine: observational studies, electronic health records, biobanks and continuous monitoring. Hum Mol Genet **27**(R1), R56–R62[30] (2018)
13. P.B. Jensen, L.J. Jensen, S. Brunak, Mining electronic health records: towards better research applications and clinical care. Nat Rev Genet **13**(6), 395–405[47] (2012)
14. R. Miotto, F. Wang, S. Wang, X. Jiang, J.T. Dudley, Deep learning for healthcare: review, opportunities and challenges. Brief Bioinformatics **19**(6), 1236–1246[72] (2018)
15. Z. Obermeyer, B. Powers, C. Vogeli, S. Mullainathan, Dissecting racial bias in an algorithm used to manage the health of populations. Science **366**(6464), 447–453[76] (2019)
16. G. Hripcsak, J.D. Duke, N.H. Shah, C.G. Reich, V. Huser, M.J. Schuemie, M.A. Suchard, R.W. Park, I.C.K. Wong, P.R. Rijnbeek, et al., Observational health data sciences and infor-

matics (ohdsi):opportunities for observational researchers. Stud Health Technol Inform **216**, 574[44] (2015)
17. S. Boughorbel, F. Jarray, N. Venugopal, S. Moosa, H. Elhadi, M. Makhlouf, Federated uncertainty-aware learning for distributed hospital ehr data. arXiv:1910.12191[9] (2019)
18. R. Duan, M.R. Boland, Z. Liu, Y. Liu, H.H. Chang, H. Xu, H. Chu, C.H. Schmid, C.B. Forrest, J.H. Holmes, et al., Learning from electronic health records across multiple sites: a communication-efficient and privacy-preserving distributed algorithm. J Am Med Inform Assoc **27**(3), 376–385[25] (2020)
19. J. Gruendner, T. Schwachhofer, P. Sippl, N. Wolf, M. Erpenbeck, C. Gulden, L.A. Kapsner, J. Zierk, S. Mate, M. Sturzl, et al., KETOS: Clinical decision support and machine learning as a service–A training and deployment platform based on Docker, OMOP-CDM, and FHIR Web Services. PloS one **14**(10), 1–16[34] (2019)
20. L. Huang, D. Liu, Patient clustering improves efficiency of federated machine learning to predict mortality and hospital stay time using distributed electronic medical records. arXiv:1903.09296[45] (2019)
21. Z. Li, K. Roberts, X. Jiang, Q. Long, Distributed learning from multiple ehr databases:Contextual embedding models for medical events. J Biomed Inform **92**, 103138[65] (2019)
22. P.V. Raja, E. Sivasankar, Modern framework for distributed healthcare data analytics based on hadoop, in *Information and communication technology-EurAsia conference*, (Springer, 2014), pp. 348–355[82]
23. J. Lee, J. Sun, F. Wang, S. Wang, C.H. Jun, X. Jiang, Privacy-preserving patient similarity learning in a federated environment: development and analysis. JMIR Medical Informatics **6**(2), e20[62] (2018)
24. Y. Kim, J. Sun, H. Yu, X. Jiang, Federated tensor factorization for computational phenotyping, in *Proceedings of the 23rd ACM SIGKDD International conference on knowledge discovery and data mining*, (ACM, 2017), pp. 887–895
25. D. Liu, D. Dligach, T. Miller, Two-stage federated phenotyping and patient representation learning. arXiv:1908.05596[67] (2019)
26. T.S. Brisimi, R. Chen, T. Mela, A. Olshevsky, I.C. Paschalidis, W. Shi, Federated learning of predictive models from federated electronic health records. Int J Med Inform **112**, 59–67[10] (2018)
27. P. Sharma, F.E. Shamout, D.A. Clifton, Preserving patient privacy while training a predictive model of in-hospital mortality. arXiv:1912.00354[90] (2019)
28. P. Vepakomma, O. Gupta, T. Swedish, R. Raskar, Split learning for health: distributed deep learning without sharing raw patient data. arXiv:1812.00564[103] (2018)
29. O. Gupta, R. Raskar, Distributed learning of deep neural network over multiple agents. J Netw Comput Appl 116, 1–8 (2018); [3] S. Silva, B. Gutman, E. Romero, P.M. Thompson, A. Altmann, M. Lorenzi, Federated learning in distributed medical databases: meta-analysis of large-scale subcortical brain data. arXiv:1810.08553[93] (2018); E. Tramel, Federated learning: rewards & challenges of distributed private ml. ccessed May 28, 2019[99] (2019)
30. S.R. Pfohl, A. M. Dai, K. Heller, Federated and differentially private learning for electronic health records (2019)
31. J. Konecny, H.B. McMahan, D. Ramage, P. Richtarik, Federated optimization: distributed machine learning for on-device intelligence. arXiv:1610.02527[57] (2016)
32. J. Konecny, B. McMahan, D. Ramage, Federated optimization: distributed optimization beyond the datacenter. arXiv:1511.03575[56] (2015)
33. B. McMahan, E. Moore, D. Ramage, S. Hampson, B.A. Arcas, Communication-efficient learning of deep networks from decentralized data, in *Artificial Intelligence and Statistics*, (2017), pp. 1273–1282[69]
34. S. Silva, B. Gutman, E. Romero, P.M. Thompson, A. Altmann, M. Lorenzi, Federated learning in distributed medical databases: meta-analysis of large-scale subcortical brain data. arXiv:1810.08553[29] (2018)

35. S. Silva, B. Gutman, E. Romero, P.M. Thompson, A. Altmann, M. Lorenzi, Federated learning in distributed medical databases: meta-analysis of large-scale subcortical brain data. arXiv:1810.08553[57] (2018)
36. R. Gellish, B. Goslin, R. Olson, A. McDonald, G. Russi, V. Moudgil, Longitudinal modeling of the relationship between age and maximal heart rate. Med. Sci. Sports Exerc. **39**, 822–829[59] (2007)
37. R.R. Singh, S. Conjeti, R. Banerjee, An approach for real-time stress-trend detection using physiological signals in wearable computing systems for automotive drivers, in *Proceedings of the 14th International IEEE Conference on Intelligent Transportation Systems*, (Washington, DC, USA, 5–7 October 2011), pp. 1477–1482.[36]
38. Y. Mao, W. Chen, Y. Chen, C. Lu, M. Kollef, T. Bailey, An integrated data mining approach to real-time clinical monitoring and deterioration warning, in *Proceedings of the 18th ACM SIGKDD International Conference on Knowledge Discovery And Data Mining*, (Beijing, China, 16–18 August 2012), pp. 1140–1148.[60]
39. D. Sow, D. Turaga, M. Schmidt, Mining of sensor data in healthcare: A survey, in *Managing and Mining Sensor Data*, ed. by C. C. Aggarwal, (Springer, Berlin, Germany, 2013), pp. 459–504.[1]
40. O.D. Lara, M.A. Labrador, A survey on ambient-assisted living tools for older adults. IEEE Commun. Surv. Tutor. **15**, 1192–1209[13] (2013)
41. J. Gialelis, P. Chondros, D. Karadimas, S. Dima, D. Serpanos, Identifying chronic disease complications utilizing state of the art data fusion methodologies and signal processing algorithms, in *Wireless Mobile Communication and Healthcare*, ed. by K. S. Nikita, J. C. Lin, D. I. Fotiadis, M. T. Arredondo Waldmeyer, vol. 83, (Springer, Berlin, Germany, 2012), pp. 256–263.[29]
42. B. Thakker, A.L. Vyas, Support vector machine for abnormal pulse classification, Int. J.Comput. Appl. **22**, 13–19.[32] (2011)
43. F. Hu, M. Jiang, L. Celentano, Y. Xiao, Robust medical ad hoc sensor networks (MASN) with Wavelet-based ECG data mining. Ad. Hoc. Netw. 6, 986–1012.[62] (2008)
44. D. Giri, U. Rajendra Acharya, R.J. Martis, S. Vinitha Sree, T.C. Lim, T. Ahamed VI, J.S. Suri, Automated diagnosis of Coronary Artery Disease affected patients using LDA, PCA,ICA and Discrete Wavelet Transform, Know. Based Syst. **37**, 274–282.[45] (2013)
45. C. Cortes, V. Vapnik, Support-vector networks. Mach. Learn. **20**, 273–297 (1995)
46. C. Bellos, A. Papadopoulos, R. Rosso, D.I. Fotiadis, A support vector machine approach for categorization of patients suffering from chronic diseases, in *Wireless Mobile Communication and Healthcare*, ed. by K. S. Nikita, J. C. Lin, D. I. Fotiadis, M. T. Arredondo Waldmeyer, vol. 83, (Springer, Berlin, Germany, 2012), pp. 264–267
47. R. Bellazzi, B. Zupan, Predictive data mining in clinical medicine: Current issues and guidelines. Int. J. Med. Inform. **77**, 81–97 (2008)
48. F. Amato, A. Lopez Rodriguez, E.M. Peña-Méndez, P. Vanhara, A. Hampl, J. Havel, Artificial neural networks in medical diagnosis. J Appl. Biomed. **11**, 47–58 (2013)
49. Q. Li, G.D. Clifford, Dynamic time warping and machine learning for signal quality assessment of pulsatile signals. Physiol. Meas. **33**, 1491–1501 (2012)
50. J.A. Lopez-Vallverdù, D. Riano, J.A. Bohada, Improving medical decision trees by combining relevant health-care criteria. Expert Syst. Appl. **39**, 11782–11791 (2012)
51. C.A. Frantzidis, C. Bratsas, M.A. Klados, E. Konstantinidis, C.D. Lithari, A.B. Vivas, C.L. Papadelis, E. Kaldoudi, C. Pappas, P.D. Bamidis, On the classification of emotional biosignals evoked while viewing affective pictures: An integrated data-mining-based approach for healthcare applications. Trans. Inf. Tech. Biomed. **14**, 309–318 (2010)
52. J.Y. Yeh, T.H. Wu, C.W. Tsao, Using data mining techniques to predict hospitalization of hemodialysis patients. Decis. Support Syst. **50**, 439–448 (2011)
53. C. Bellos, A. Papadopoulos, R. Rosso, D.I. Fotiadis, Categorization of patients' health status in copd disease using a wearable platform and random forests methodology, in *Proceedings of*

the IEEE International Conference on Biomedical and Health Informatics, (Shenzhen, China, 5–7 January 2012), pp. 404–407
54. F.T. Sun, C. Kuo, H.T. Cheng, S. Buthpitiya, P. Collins, M. Griss, Activity-aware mental stress detection using physiological sensors, in *Mobile Computing, Applications, and Services*, ed. by M. Gris, G. Yang, vol. 76, (Springer, Berlin, Germany, 2012), pp. 211–230
55. J.Y. Yeh, T.H. Wu, C.W. Tsao, Using data mining techniques to predict hospitalization of hemodialysis patients. Decis. Support Syst. **50**, 439–448 (2011)
56. W. Wang, H. Wang, M. Hempel, D. Peng, H. Sharif, H.H. Chen, Secure stochastic ECG signals based on the gaussian mixture model for e-healthcare systems. IEEE Syst. J. **5**, 564–573 (2011)
57. B.S. Bhati, C.S. Rai, Analysis of support vector machine-based intrusion detection techniques. Arabian J. Sci. Eng. **45**(4), 2371–2383 (2020). https://doi.org/10.1007/s13369-019-03970-z
58. J. Kalagnanam, M. Henrion, A comparison of decision analysis and expert rules for sequential diagnosis. arXiv:1304.2362 (2013)
59. F.T. Sun, C. Kuo, H.T. Cheng, S. Buthpitiya, P. Collins, M. Griss, Activity-aware mental stress detection using physiological sensors, in *Mobile Computing, Applications, and Services*, ed. by M. Gris, G. Yang, vol. 76, (Springer, Berlin, Germany, 2012), pp. 211–230
60. K.V.P. Naraharisetti, M. Bawa, M. Tahernezhadi, Comparison of different signal processing methods for reducing artifacts from photoplethysmograph signal, in *Proceedings of the IEEE International Conference on Electro/Information Technology*, (Mankato, MN, USA, 15–17 May 2011), pp. 1–8
61. D. Apiletti, E. Baralis, G. Bruno, T. Cerquitelli, Real-time analysis of physiological data to support medical applications. Trans. Info. Tech. Biomed. **13**, 313–321 (2009)
62. G.N. Pradhan, R. Chattopadhyay, S. Panchanathan, Processing body sensor data streams for continuous physiological monitoring, in *Proceedings of the International Conference on Multimedia Information Retrieval*, (Philadelphia, PA, USA, 29–31 March 2010), pp. 479–486
63. O. Salem, Y. Liu, A. Mehaoua, A lightweight anomaly detection framework for edical wireless sensor networks, in *Proceedings of the IEEE Wireless Communications and Networking Conference*, (Shanghai, China, 7–10 April 2013), pp. 4358–4363
64. L. Rabiner, B.H. Juang, An introduction to hidden Markov models. IEEE ASSP Mag. **3**, 4–16 (1986)
65. A. Hard, K. Rao, R. Mathews, F. Beaufays, S. Augenstein, H. Eichner, C. Kiddon, D. Ramage, Federated learning for mobile keyboard prediction. arXiv:1811.03604 (2018)
66. S. Samarakoon, M. Bennis, W. Saad, M. Debbah, Federated learning for ultra-reliable low-latency v2v communications, in *2018 IEEE Global Communications Conference (GLOBECOM)*, (IEEE, 2018), pp. 1–7

Index

S. P. Yadav et al. (eds.), *Federated Learning for IoT Applications*, EAI/Springer Innovations in Communication and Computing,
https://doi.org/10.1007/978-3-030-85559-8

FSC
www.fsc.org
MIX
Papier aus verantwortungsvollen Quellen
Paper from responsible sources
FSC® C105338

If you have any concerns about our products,
you can contact us on
ProductSafety@springernature.com

In case Publisher is established outside the EU,
the EU authorized representative is:
Springer Nature Customer Service Center GmbH
Europaplatz 3, 69115 Heidelberg, Germany

Printed by Libri Plureos GmbH
in Hamburg, Germany